AF658807

HORMONES, BRAIN AND BEHAVIOUR OF FISH

Published By

Oxford Book Company

Regd. Off. "Bony Residency", Gate No. 2
Opp. Tilak Public School, Vishweswariya Nagar,
Gopalpura Road, Jaipur - 302018, India
Ph: 0141 - 2761280, Telefax: 2761381
email - oxfordbookjpr@gmail.com
website - oxfordbookcompany.com

Delhi Office: 102, 1st Floor, Satyam House,
4327/3, Ansari Road, Darya Ganj,
New Delhi - 110002, India
Ph: 011-45652440

Distributed By
KEYA INTERNATIONAL PRESS LLP
E-mail: keyainternationalpressllp@gmail.com

Published in India, 2025

© All Rights Reserved

ISBN : 9789391096410

All Rights Reserved. No part of this publication may be reproduced, stored in a retrieval system, or transmitted, in any form or by any means, electronic, mechanical, photocopying, recording, scanning or otherwise, without the prior written permission of the copyright owner.

Typeset by Shivangi Computers, Jaipur

Printed and bound in India by Thomson Press Ltd.,
New Delhi

Contents

Preface

Aggression may be defined as overt behaviour with the intention of inflicting harm or the threat of harm upon another individual. Aggressive behaviours range from lethal to subtle. The possibility for aggressive behaviour exists whenever the interests of two or more individuals conflict. Conflicts typically arise over limited resources, including territories, mates, and food. Among social species, conflicts may arise over social status and limited resources. A social interaction decides which individual gains status or access to the contested resource. In many cases, a submissive posture or gesture from one individual avoids actual combat. Individuals may also participate in threat displays or ritualized combat in which dominance is determined, but no physical damage is inflicted. Species- and situation-specific rules exist to regulate aggression.

When aggressive behaviours break these rules or when the aggression is excessive to situational norms, then it may be considered to be pathological or violent. Pathological or violent aggression causes much personal and societal suffering, and thus is an important topic for study. Treatment for violent or pathological aggressive behaviour remains primitive. Essentially, no effective interventions exist for violent humans – the most common treatment for violent people is incarceration. Remarkably, a death sentence is the sole treatment for an aggressive dog that bites a person. Thus, understanding the pathophysiology of excessive aggression remains an important, yet understudied endeavour. One factor limiting progress is the validity of animal models of human violent behaviour. For example, in studies of animal models of aggression, it is rare that normal, adaptive aggression is differentiated from pathological violence. Indeed, in common with categorizing human behaviour, it is often difficult for observers to draw the line between adaptive aggression and pathological aggression in nonhuman animals.

Thus, the neural and biological mechanisms underlying violent behaviour are not well understood, although our understanding of the mechanisms underlying aggressive behaviour is growing. This book focuses on the interaction between hormones and aggressive behaviours. For the purposes of this book, two broad associations suggest that hormones are involved in aggression. First, a dramatic sex difference exists in aggression and violence across species and situations: males are typically more aggressive than females.

A notable exception to this bromide is that females defending their young are fiercely aggressive. The extent to which human sex differences in violence are due to hormones (or other physiological differences) or reflect differences in socialization remains unspecified. Moderate testosterone concentrations are necessary for expression of male aggression in most strains of mice and perhaps humans as well. Second, the emergence of aggression and violence typically occurs at puberty coinciding with increased secretion of sex steroid concentrations. Within the past three decades, a better appreciation of the broad endocrine correlates to aggressive behaviours has been obtained, but specifics have only recently emerged.

In addition to precision in assaying hormones, additional insights into hormone–aggressive behaviour interactions require precise behavioural analyses. Obtaining blood samples from individuals engaged in an aggressive encounter may not provide useful endocrine correlates without detailed behavioural analyses, or identification of the social status of both participants. The term agonistic is used to describe the complete behavioural repertoire of both aggressive and submissive responses within the context of a social interaction involving a conflict of interest. However, submissive behaviours are not aggressive behaviours; consequently, hormonal mediation of the aggressive and the submissive components of an agonistic interaction likely differ. Hormones are closely associated with the regulation of aggressive behaviour across species, but due to lack of space, this book focuses on mammals, primarily rodents and primates.

The book will not only meet the requirement of students but will also be useful as a guideline to the teachers and researchers.

— Dipti Nagar

Chapter 1

Fish Hormones

Anti-Müllerian Hormone

Anti-Müllerian hormone also known as AMH is a protein that, in humans, is encoded by the *AMH* gene. It inhibits the development of the Müllerian ducts (paramesonephric ducts) in the male embryo. It has also been called Müllerian inhibiting factor (MIF), Müllerian-inhibiting hormone (MIH), and Müllerian-inhibiting substance (MIS). It is named after Johannes Peter Müller.

Species Distribution: AMH is present in fish, reptiles, birds, marsupials, and placental mammals.

Source: AMH is secreted by Sertoli cells of the testes during embryogenesis of the fetal male.

Structure: AMH is a protein hormone structurally related to inhibin and activin, and a member of the transforming growth factor-â (TGF-â) family. It is a dimeric glycoprotein.

Gene: In humans, the gene for AMH is *AMH*, on chromosome 19p13.3,[1] while the gene *AMHR2* codes for its receptor on chromosome 12.

Function

Embryogenesis: In mammals, AMH prevents the development of the mullerian ducts into the uterus and other mullerian structures. The effect is ipsilateral, that is each testis suppresses Müllerian development only on its own side. In humans, this action takes place during the first 8 weeks of gestation. If no hormone is produced from the gonads, the Mullerian ducts automatically develop, while the Wolffian ducts, which are responsible for male reproductive ducts, automatically die. Amounts of AMH that are measurable in the blood vary by age and sex. AMH works by interacting with specific receptors on the surfaces of the cells of target tissues. The best-known and most specific effect, mediated through the AMH type II receptors, includes programmed cell death (apoptosis) of the target tissue (the fetal mullerian ducts).

Ovarian

In healthy females AMH is either just detectable or undetectable in cord blood at birth and demonstrates a marked rise by three months of age; while still detectable it falls until four years of age before rising linearly until eight years of age remaining fairly constant from mid-childhood to early adulthood - it does not change significantly during puberty; from 25 years of age AMH declines to undetectable levels at menopause. AMH is expressed by granulosa cells of the ovary during the reproductive years, and controls the formation of primary follicles by inhibiting excessive follicular recruitment by FSH. It, therefore, has a role in folliculogenesis, and some authorities suggest it is a measure of certain aspects of ovarian function, useful in assessing conditions such as polycystic ovary syndrome and premature ovarian failure.

Other

AMH production by the Sertoli cells of the testes remains high throughout childhood in males but declines to low levels during puberty and adult life. AMH has been shown to regulate production of sex hormones, and changing AMH levels (falling in females, rising in males) may be involved in the onset of puberty in both sexes. Functional AMH receptors have also been found to be expressed on neurons in the brains of embryonic mice, and are thought to play a role in sexually dimorphic brain development and consequent development of gender-specific behaviours.

Pathology

In men, inadequate embryonal AMH activity can lead to the *Persistent Müllerian duct syndrome* (PMDS), in which a rudimentary uterus is present and testes are usually undescended. The AMH gene (*AMH*) or the gene (*AMH-RII*) for its receptor are usually abnormal. AMH measurements have also become widely used in the evaluation of testicular presence and function in infants with intersex conditions, ambiguous genitalia, and cryptorchidism.

Application

AMH has been synthesized. Its ability to inhibit growth of tissue derived from the Müllerian ducts has raised hopes of usefulness in the treatment of a variety of medical conditions including endometriosis, adenomyosis, and uterine cancer. Research is underway in several laboratories. Comparison of an individual's AMH level with respect to average levels is also useful in fertility assessment, as it provides a guide to ovarian reserve and identifies women that may need to consider either egg freezing or trying for a pregnancy sooner rather than later if their long-term future fertility is poor. Measuring AMH alone may be misleading as high levels occur in conditions like polycystic ovarian syndrome and therefore AMH levels should be considered in conjunction with a transvaginal scan of the ovaries to assess antral follicle count and ovarian volume.

It also has the potential to rationalise the programme of ovulation induction and decisions about the number of embryos to transfer in assisted reproduction techniques to maximise pregnancy success rates whilst minimising the risk of ovarian hyperstimulation syndrome (OHSS) AMH can predict an excessive response in ovarian hyperstimulation with a sensitivity and specificity of 82% and 76%, respectively.

Honeycomb Grouper

The Honeycomb grouper (*Epinephelus merra*) is one of the smaller fish species in the Epinephelus genus, with a maximum known length of 31 cm. *E. merra* typically feeds on crustaceans and fish. E. merra is commonly found in reef environments of the Pacific Ocean, from South Africa to the areas around Australia, Japan and the surrounding islands. The honeycomb grouper is a valuable food fish. Attempts are underway to culture this species in farms. However males are rare, which limits commercial production of this fish.

Hermaphroditism

The genus *Epinephelus* has the ability to change sexes, more specifically they are protogynous hermaphrodites what means females can turn into males at one point in their life. The Honeycomb grouper is studied because of its relatively small size. This sex change usually happens in the non breeding season and can be caused by social stimuli, age, growth, and body size. Yet, the initial trigger for the sex change remains unknown. Females turn into males when they reach a length of approximately 20 cm, thus the size advantage model for sequential hermaphroditism applies in the case of the honeycomb grouper. The size advantage model states that if an individual can reproduce more effectively as one sex when small or young and as the other sex when larger or older, it should change sex at some point in its life history.

In summary, the Honeygrouper has been found to transition from female to male when low levels of estradiol-17â (E2) exists, leading to increased levels of 11-ketotestosterone (11KT). These hormonal changes lead to a degeneration of existing oocytes and trigger the sex change. When E2 levels fall below a specific physiological threshold, oocytes no longer survive and spontaneous androgen levels now initiate the proliferation of sperm producing cells. High levels of 11KT and testosterone maintain spermatogenesis.

Estrogen's (E2) role: Females only change sex after meeting or exceeding a length of 20 cm and during the nonbreeding season, where E2 has been found to be low. Yet it is still unknown if the drop in E2 is the cause or an effect of the sex change. In fish, it is conclusive that estrogen has a direct effect on ovarian differentiation and is needed to maintain overall ovarian development. Biosynthesis of estrogen is mediated by aromatase P450.

Artificial Aromatase Inhibitors (AI): When an (AI) is administered, the levels of Estrogen decrease and even immature females turn into males. However, if AI is given together with artificial E2 the fish remains female. This finding concludes that E2 plays a big role in maintaining a female gender in the Honeycomb grouper fish. Lack of Estrogen causes degradation of oocytes, and the start of sperm production. AI could be a tool to increase the breeding rates of grouper fish, by increasing both sperm count and fertility rates in the fish.

11-ketotestosterone's (11KT) role: 11-KT's levels increase in the fish in correlation with the transition from female to male. When 11KT was induced it triggered a sex change. 11KT stimulates the growth of male germ cells and their differentiation.

Stages of Transition

Ovarian: Immature female, resting female, developing female, maturing female, mature female, partially running female, and post-spawning female.

Transitional: Early transitional, bisexual and late transitional.

Testicular: Maturing, mature and running ripe testes

Stages are determined by proportion between male and female tissue, female germ cell degradation, presence of male cysts arranged in lobules, and development of sperm sinuses along gonadal wall.

Salcatonin

Salcatonin is the type of calcitonin hormone found in salmon. As in humans, salmon calcitonin is a peptide hormone secreted by the parafollicular cells of the thyroid gland in response to hypercalcemia, which lowers blood calcium and phosphate by promoting renal excretion. Synthetic salmon calcitonin may be used therapeutically in humans, as it is twenty times more active than human calcitonin and has a longer half-life. It is used as therapy for Paget's disease and severe hypercalcemia.

Vitellogenin

Vitellogenin (VTG or less popularly known as VG) (from latin vitellus = yolk and gener = to produce) is a synonymous term for the gene and the expressed protein. The protein molecule is classified as a glycolipoprotein, having properties of a sugar, fat and protein. It belongs to a family of several lipid transport proteins.

Vitellogenin is an egg yolk precursor protein expressed in the females of nearly all oviparous species including fish, amphibians, reptiles, birds, most invertebrates, and the platypus. Vitellogenin is the precursor of the lipoproteins and phosphoproteins that make up most of the protein content of yolk. In the presence of estrogenic endocrine disruptive chemicals (EDCs), male fish can express the Vg gene in a dose dependent manner. Vg gene expression in male fish can be used as a molecular marker of exposure to estrogenic EDCs.

Vitellogenin and Honey Bees

Honey bees deposit vitellogenin molecules in fat bodies in their abdomen and heads. The fat bodies apparently acts as a food storage reservoir. The glycolipoprotein vitellogenin has additional functionality as it acts as an antioxidant to prolong Queen bee and forager lifespan as well as a hormone that affects future foraging behaviour. The health of a honey bee colony is dependent upon the vitellogenin reserves of the nurse bees - the foragers having low levels of vitellogenin. As expendable labourers, the foragers are fed just enough protein to keep them working their risky task of collecting nectar and pollen. Vitellogenin levels are important during the nest stage and thus influence honey bee worker division of labour.

A nurse bee's vitellogenin titer that developed in the first four days after emergence, affects its subsequent age to begin foraging and whether it preferentially forages for nectar or pollen. If young workers are short on food their first days of life, they tend to begin foraging early and preferentially for nectar. If they are moderately fed, they forage at normal age preferentially for nectar. If they are abundantly fed, immediately after emergence, their vitellogenin titer is high and they begin foraging later in life, preferentially collecting pollen. Pollen is the only available protein source for honey bees.

Vitellogenin and Juvenile Hormone Feedback Loop

For the majority of the investigated insect species it has been documented that juvenile hormone stimulates the transcription of the vitellogenin genes and the consequent control of vitellogenin production (cf. Engelmann, 1983; Wyatt and Davey, 1996). The vitellogenin expression is part of a regulatory feedback loop that enables vitellogenin and juvenile hormone to mutually suppress each other. Vitellogenin and juvenile hormone likely work antagonistically in the honey bee to regulate the honey bees development and behaviour. Suppression of one leads to high titers of the other.

It is likely that the balance between vitellogenin and juvenile hormone levels is also involved in swarming behaviour. Juvenile hormone levels drop in honey bee colonies pre-swarming and it is expected that vitellogenin levels would therefore rise. One may surmise, that swarming bees would want to pack along as much vitellogenin as possible to extend their lifespan and to be able to quickly build a new nest.

Induced Reproduction in Fish

Among the most significant advancements in the field of aquaculture during recent decades is the development of techniques to induce reproduction in fish. These techniques have allowed farmers to profitably breed and raise species that do not naturally reproduce in captivity, and to manipulate the timing of reproduction to suit production cycles. Some species will not readily breed in captivity due to environmental or culture conditions that are different from those found in nature, such as water temperature or substrate type. These conditions may cause stress or may not provide the cues needed to complete the reproductive process.

Fish in captivity may not always reproduce at the most advantageous time, and alteration of the spawning cycle may be desirable. This allows a farmer to:

1. obtain fish outside of the normal spawning season either to lengthen time for grow-out or to produce hybrids with other species;
2. improve efficiency by getting fish to spawn on a predetermined date; and

3. maximise survival by fertilizing and incubating eggs under hatchery conditions. Where successful, techniques for altering the spawning cycle of fish have become a valuable tool.

Background: Reproduction in Nature

In fish, the reproductive process involves three basic steps:

1. maturation - the development of the gametes (eggs and sperm) to a point where fertilization can occur;
2. ovulation - the release of eggs from the ovary; and
3. spawning - the deposition of eggs and sperm so that they can unite.

In fish, as with all higher animals, hormones play a critical role in the reproductive process. Hormones are chemical messengers released into the blood by specific tissues, such as the pituitary gland. The hormones travel through the bloodstream to other tissues, which respond in a variety of ways. One response is to release another hormone, which elicits a response in yet another tissue. The primary tissues involved in this hormonal cascade are the hypothalamus, pituitary gland, and gonads. Fish have evolved to reproduce under environmental conditions that are favourable to the survival of the young. Long before spawning, seasonal cues begin the process of maturation.

In many fish, this can take up to a year. When the gametes have matured, an environmental stimulus may signal the arrival of optimal conditions for the fry, triggering ovulation and spawning. Examples of environmental stimuli are changes in photoperiod, temperature, rainfall, and food availability. A variety of sensory receptors detect these cues, including the eye, pineal gland (an organ in the dorsal part of the forebrain that is sensitive to light), olfactory organs, taste buds, and thermoreceptors.

The hypothalamus, located at the base of the brain, is sensitive to signals from sensory receptors and releases hormones in response to environmental cues. Principal among these hormones are gonadotropin releasing hormones (GnRH), which travel from the hypothalamus to the pituitary gland. The pituitary is responsible for a wide variety of functions, including growth and reproduction. Certain cells of the pituitary receive GnRH and release gonadotropic hormones into the bloodstream. The gonadotropic hormones travel to the gonads, which synthesize steroids responsible for final maturation of the gametes.

Maturation of the egg is a long process that involves complex physiological and biochemical changes. One important step, vitellogenesis, is a process in which yolk proteins are produced in the liver, transported to the ovary, and stored in the egg, resulting in tremendous egg enlargement. The yolk is important as a source of nutrition for the developing embryo.

Also critical are germinal vesicle migration and germinal vesicle breakdown (GVBD). Before it migrates, the germinal vesicle, or nucleus, is located at the centre of the egg in an arrested stage of development. At this stage, the egg is physiologically and genetically incapable of being fertilized, even though it has the outward appearance of a fully mature egg. When conditions are appropriate for final maturation, nuclear development resumes, and the germinal vesicle migrates to one side. Finally, the walls of the germinal vesicle break down, releasing the chromosomes into the cell.

The maturity of eggs can be determined using biopsy techniques. Eggs are removed from the ovaries, cleared with a prepared solution, and viewed under a microscope. In mature eggs, the migration of the germinal vesicle to the side of the cell will be complete.

After the egg has matured, a class of compounds called prostaglandins are synthesized. These stimulate ovulation, which is the rupture of the follicle cells that hold the egg. The egg is then released into the body cavity or ovarian lumen, where it may subsequently be released to the outside environment. Following ovulation, the viability of the eggs can decrease rapidly.

Fish with gametes that have not yet been released by the gonads are called "green." The term "ripe" is used to describe fish with gametes that have been released from the ovary into the ovarian lumen. Ripe fish can be stripped, green fish cannot.

Inducing Reproduction

There are two main strategies used to induce reproduction. The first is to provide an environment similar to that in which spawning occurs naturally. Catfish, for example, like to spawn in enclosed spaces such as hollow logs. A farmer can simulate this by putting milk cans in a pond. The presence of vegetation and an increase in temperature will usually work for goldfish. Changing the photoperiod in a hatchery can accelerate or delay maturation and ovulation in many salmon and trout species.

The second strategy is to inject the fish with one or more naturally occurring reproductive hormones or their synthetic analogs. This is only effective in fish that are already in breeding condition and have mature eggs in which the germinal vesicle has migrated. Often the two strategies are used sequentially: the first to manipulate maturation, then the second to induce ovulation. Numerous hormones have been used to induce reproduction. Two methods have emerged over the past few years that seem to offer the best chance for success at the least expense. They are injection of a GnRH analog with dopamine antagonist, and injection of gonadotropin.

GNRH Analogue with Dopamine Antagonist

Leutinizing Hormone Releasing Hormone (LHRH) is the name of a mammalian hormone that has been employed successfully to induce the reproductive hormonal cascade. In recent years, synthetic analogues of LHRH, referred to as LHRHa, have been developed that are far more effective. Because they are purer and are not rapidly metabolized by fish, they remain active for longer periods. Under natural conditions, there is a feedback mechanism in the fish that limits the release of gonadotropin. This mechanism uses a chemical called dopamine, which inhibits the action of LHRH. When dopamine is present in the fish, even LHRHa will have only limited success. A dopamine antagonist is often used to limit the effects of dopamine. When LHRHa and a dopamine antagonist are used in conjunction, reproductive success dramatically increases.

Gonadotropin

Two types of gonadotropin extracts have been used to induce ovulation in fish: Human Chorionic Gonadotropin (HCG) and fish pituitary extract. Pituitary extracts are made by removing the pituitary from a fish and extracting the hormones, which may then be injected into another fish. Carp, catfish, salmon, and other fish have been used for this. HCG offers three major advantages over the pituitary extract: 1) it is much less expensive, 2) it is more stable and thus has a longer shelf life, and 3) it comes in a purified form.

Methods

The details of inducing spawning differ from species to species. They will also differ according to the goals and means of the farmer. It must be emphasized that the following techniques do not apply to all situations. Farmers should always thoroughly research the techniques that have been developed for their species of fish and select those that best fit the circumstances.

Broodstock

Brooders should be fast growing, disease free, and sexually ripe. There are several indicators of ripeness:

Females

- The abdomen is rounded and soft.
- The genital opening is swollen, protruding, and reddish.
- The anus is often also swollen and reddish.
- Secondary sexual characteristics are evident.

Males

- Milt is released when the abdomen is pressed gently.
- Secondary sexual characteristics are evident.

Brooders must be handled with extreme gentleness. They are physically very vulnerable during spawning and may die if dropped or roughly handled. Often, a good deal of the farmer's resources has gone into developing the broodstock line and keeping the fish healthy. The potential return from this investment can easily be eliminated through sloppy handling. When the brooders are kept in ponds or pens, all water quality parameters, such as temperature and oxygen, should be kept at optimal levels. A satisfactory amount of high quality, well balanced feed should be provided. If the fish are moved from one set of conditions to another, time must be allowed for acclimation. When handling the fish, gentle firmness should be the rule. Covering the fish with wet burlap or cloth, or covering their container to cut off light, will help to keep them calm. Tranquilizers may be used when fish are to be injected or stripped.

Injecting the Fish

There are two common places to inject hormones into a fish. An intraperitoneal (within the body cavity) injection is given through the ventral (bottom) part of the fish behind either the pelvic or pectoral fin. Intramuscular (within the muscle) injections are commonly done on the dorsal (upper) part of the fish above the lateral line and below the anterior part of the dorsal fin. In either case, it is important to place the needle so that it slides under the scale rather than through it.

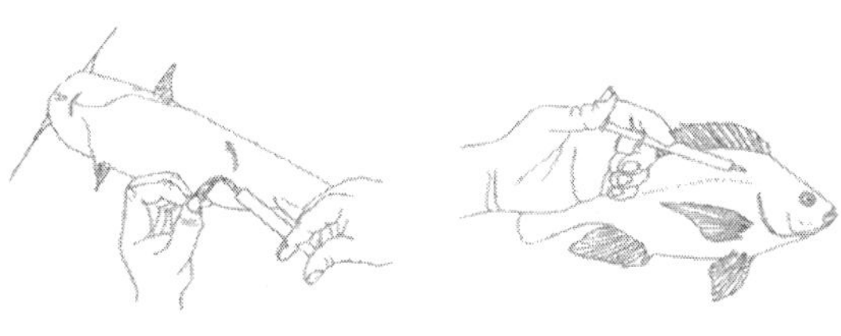

Two dosage levels are commonly used: a preparatory dose and a decisive, or final, dose with a time gap generally of 12 to 24 hours between the two injections. The preparatory dose brings the fish to the brink of spawning, and the decisive dose induces ovulation. In general, the preparatory dose is about 10 percent of the total dose. For some fish, several preparatory doses may be necessary.

Stripping

Stripping should be carried out as soon after ovulation as possible. Fish should be anesthetized and examined 6 to 12 hours after the final injection. Before stripping, both the male and female should be cleaned and dried around the vent with a soft towel because residual anesthetic will kill sperm. Water, fish "slime", feces, and blood should not be allowed to mix with the gametes as they are stripped.

To strip the fish, the female should be held around the caudal fin with one hand, while applying slight pressure to the abdomen with the other hand. If ovulation has occurred, a stream of eggs will emerge. If only a few appear, the female is still "green" and should be returned to the holding tank. If there is a stream of eggs, the abdomen should be messaged from front to back to strip out all the eggs.

Examples

The following procedures provide examples of a variety of techniques used to induce spawning.

Golden Shiners

Some fish can be induced to spawn simply by providing an appropriate environment. An example is furnishing spawning substrate for golden shiners, which naturally lay their eggs on vegetation. There are two common ways of doing this. The first is to plant natural vegetation around the shore. The second is to place spawning mats in the pond. The advantage of spawning mats is that once the eggs are laid, the mats can be removed and placed in another pond for grow-out. One common method for constructing these mats is to sandwich Spanish moss or a similar material inside steel-welded wire. A piece of 2" by 4" steel mesh is folded in half so that the dimensions of the mat are about 20" by 40". The Spanish moss is placed inside the fold, and hognose rings are used to hold the bottom together. The following is adapted from *Third Report to the Fish Farmers*, published by the United States Fish and Wildlife Service.

1. Twenty to forty pounds of brood stock may be stocked per acre.
2. Plankton blooms can inhibit spawning. If heavy blooms occur, the pond should be flushed with fresh water.
3. If grass is to be planted, the water level in the brood pond should be lowered during early spring to allow growth along the shoreline. Once the grass has been established, the ponds should be refilled.
4. If spawning mats are to be used, they should be placed in shallow water, about one inch below the surface with one side at the edge of the pond. Several mats may be placed end-to-end.
5. As each mat fills with eggs, it should be transferred to a rearing pond for grow-out. Care must be taken so that the mats do not become over-populated with eggs, because they may eventually smother each other or be attacked by fungus.
6. If spawning activity diminishes quickly, a rapid rise in the water level may extend activity.

Trout

In nature, trout build their nests in gravel and require cold, flowing, aerated water to reproduce. Changes in photoperiod are critical for bringing the fish to spawning condition. The spawning season varies with locality, temperature, and the species or strain of fish. Brook and brown trout are fall spawners, while rainbow trout spawn in the spring. The following steps, summarised from Fish Culture Manual, published by the Alaska Department of Fish and Game, illustrate one method for spawning trout by stripping females and mixing the eggs with milt. Many other methods also exist.

1. Anesthetize the fish. Placing them in MS-222 at a concentration of 130-260 mg/l (assuming temperature between 4¡C and 15¡C) is common. The solution should be within 5¡C of the holding or rearing containers, and the pH should be between 6 and 8.
2. Holding the caudal peduncle with one hand and supporting the head with the other, dip the fish in clean water to wash off the anesthetic.
3. Strip the female's eggs into a dry, shallow container. The fish should be held belly down in a horizontal position with the head slightly higher than the tail. Gently press upward near the base of the pelvic fin to start the flow of eggs. Slowly stroke the belly, moving the hand from front to rear using constant pressure. Repeat this procedure until no more eggs are obtained.

4. In the dry method of fertilization, sperm from the males is poured onto the eggs. Water should not come into contact with the mixture for it will inhibit fertilization by the sperm. The sperm, eggs, and ovarian fluid should be gently mixed with a feather.
5. After 30 seconds, fertilization will have taken place. Add enough water to cover the eggs and mix again.
6. Let the fertilized eggs soak for 30 seconds.
7. Rinse the fertilized eggs in clean water.
8. Water harden the eggs. This is a process whereby the eggs absorb water and grow about 20 percent. Place the eggs into a container that will accommodate them as they grow. Add water, then do not disturb for an hour.
9. Put into incubators.

Catfish

Temperature is a determining factor in the control of catfish reproduction. Spawning activity begins after water temperatures stay consistently above 70¡F. It drops off when temperatures fall below 70¡F or rise above 85¡F. There are a variety of methods commonly used for spawning catfish, including spawning in ponds, spawning in pens, and hand stripping. In many, though not all cases, hormone injection may be desirable. The following procedure from Third Report to the Fish Farmers provides an example of spawning induction with hormones followed by hand stripping.

1. Anesthetize the fish with MS-222 at a concentration of 80 mg/l.
2. If the fish are kept in a holding tank, loosely stitch their mouths closed to prevent them from biting one another. The thread must be loose enough to allow respiration.
3. Using an 18-gauge needle, inject intraperitoneally with: a. 3.5-4.0 mg of acetone-dried pituitary gland extract per kilogram of fish for the female and about 2 mg per kilogram for the male, or b. 500-800 international units of human chorionic gonadotropin (HCG) per pound of fish.
4. If necessary, inject again 48 hours later. Usually, the female ovulates within 16-20 hours after an injection.
5. Strip the eggs and the milt into a 0.3 percent solution of common salt.
6. Very gently mix the eggs and milt. Fertilization should occur in 2-5 minutes.

7. Place the egg mass in a suitable incubator for hatching.
8. After 10-12 hours, stop the water flow and treat with a 0.3-0.5 percent solution of alkaline protease enzyme for 2-3 minutes while gently stirring. This will dissolve the sticky layer of the eggs and allow them to float free.
9. When the eggs begin hatching, they should be taken out and placed in a flat tray.

Many species of fish will not readily reproduce under certain culture conditions. Others will, but not necessarily when the farmer desires. In these cases, induction of spawning can be of great value. Two techniques are commonly used, sometimes in conjunction with one another. The first is manipulation of the culture environment to mimic some important quality in the fish's natural environment. The second is injection of hormones to stimulate spawning. The hormones may be natural hormones taken from fish or other animals, genetically engineered from bacteria, or synthetic analogs of naturally-occurring hormones.

Methods vary from species to species and situation to situation. However, at least two generalisations can be drawn. First, brooders are very vulnerable to rough handling. Care should always be used to avoid damaging these valuable animals. Second, a fish that does not have mature gametes will not produce viable eggs or sperm no matter how many times it is injected with hormones. Ripeness is the result of environmental factors working over a period of time, leading to maturation of the gonads and production of viable eggs.

Many procedures have been developed for inducing fish to undergo the last steps of spawning. Farmers should thoroughly research the procedures that have been developed for their species of fish through experimentation, and select those that best suit the circumstances. In addition, once the fish have spawned, there are many techniques involved in incubating and caring for the eggs, and caring for the hatched fry. These too must be thoroughly researched.

Biotechnology and its Applications in Aquaculture and Fisheries

Biotechnology provides powerful tools for the sustainable development of aquaculture, fisheries, as well as the food industry. Increased public demand for seafood and decreasing natural marine habitats have encouraged scientists to study ways that biotechnology can increase the production of marine food products, and making aquaculture as a growing field of animal research. Biotechnology allows scientists to identify and combine traits in fish and shellfish to increase productivity and improve quality. Scientists are investigating genes that will increase production of natural fish growth factors as well as the natural defence compounds marine organisms use to fight microbial infections. Modern biotechnology is already making important contributions and poses significant challenges to aquaculture and fisheries development. It perceives that modern biotechnologies should be used as adjuncts to and not as substitutes for conventional technologies in solving problems, and that their application should be need-driven rather than technology-driven.

The use of modern biotechnology to enhance production of aquatic species holds great potential not only to meet demand but also to improve aquaculture. Genetic modification and biotechnology also holds tremendous potential to improve the quality and quantity of fish reared in aquaculture. There is a growing demand for aquaculture; biotechnology can help to meet this demand. As with all biotech-enhanced foods, aquaculture will be strictly regulated before approved for market. Biotech aquaculture also offers environmental benefits. When appropriately integrated with other technologies for the production of food, agricultural products and services, biotechnology can be of significant assistance in meeting the needs of an expanding and increasingly urbanized population in the next millennium. Successful development and application of biotechnology are possible only when a broad research and knowledge base in the biology, variation, breeding, agronomy, physiology, pathology, biochemistry and genetics of the manipulated organism exists. Benefits offered by the new technologies cannot be fulfilled without a continued commitment to basic research. Biotechnological programmes must be fully integrated into a research background and cannot be taken out of context if they are to succeed.

Indian fisheries and aquaculture is an important sector of food production, providing nutritional security to the food basket, contributing to the agricultural exports and engaging about fourteen million people in different activities. With diverse resources ranging from deep seas to lakes in the mountains and more than 10% of the global biodiversity in terms of fish and shellfish species, the country has shown continuous and sustained increments in fish production since independence. Constituting about 4.4% of the global fish production, the sector contributes to 1.1% of the GDP and 4.7% of the agricultural GDP. The total fish production of 6.57 million metric tonnes presently has nearly 55% contribution from the inland sector and nearly the same from culture fisheries. Fish and fish products have presently emerged as the largest group in agricultural exports of India.(Marine products export review-MPEDA., April 2006-March2007). The potential area of biotechnology in aquaculture include the use of synthetic hormones in induced breeding, transgenic fish, gene banking, uniparental and polyploidy population and health management.

Biotechnology in Fish Breeding

Gonadotropin releasing hormone (GnRH) is now the best available biotechnological tool for the induced breeding of fish. GnRH is the key regulator and central initiator of reproductive cascade in all vertebrates (Bhattacharya et al.,2002). It is a decapeptide and was first isolated from pig and ship hypothalami with the ability to induce pituitary release of luteinising hormone (LH) and follicle stimulating hormone (FSH) (Schally et al.,1973). Since then only one form of GnRH has been identified in most placental mammals including human beings as the sole neuropeptide causing the release of LH and FSH. However, in non mammalian species (except guinea pig) twelve GnRH variants have now been structurally elucidated, among them seven or eight different forms have been isolated from fish species. The induced breeding of fish is now successfully achieved by development of GnRH technology.

Transgenesis

Transgenesis or transgenics may be defined as the introduction of exogenous gene / DNA into host genome resulting in its stable maintenance, transmission and expression. The technology offers an excellent opportunity for modifying or improving the genetic traits of commercially important fishers, mollusks and crustaceans for aquaculture. The idea of producting transgenic animals became popular when Palmitter et al. (1982) first produced transgenic mouse by introducing metallothionein human growth hormone fusion gene (mT-hGH) into mouse egg, resulting in dramatic increase in growth. This triggered a series of attemptson gene transfer in economically important animals including fish.

The first transgenic fish was produced Zhu et al. (1985) in China, who claimed the transient expression n putative transgenics, although they gave no molecular evidence for the integration of the transgene. The technique has now seen successfully applied to a number of fish species. Dramatic growth enhancement has been shown using this technique especially in salmonids (Devlin et al., 1994). Some studies have revealed enhancement of growth in adult salmon to an average of 3 Ð 5 times the size of non Ð transgenic controls, with some individuals, especially during the first few months of growth, reaching as much as 10 Ð 30 times the size of the controls (Devlin et al., 1994; Hew et al., 1995).The introduction of transgenic technique has simultaneously put more emphasis on the need for production of sterile progeny in order to minimise the risk of transgenic stocks mixing in the wild populations. The technical development has expanded the possibilities for producing either sterile fish or those whose reproductive activity can be specifically turned on or off using inducible promoters. This would clearly be of considerable value allowing both optimal growth and controlled reproduction of the transgenic stocks while ensuring that any escaped fish would be unable to breed. An increased resistance of fish to cold temperatures has been another subject of research in fish transgenics for the past several years (Fletcher et al., 2001). Coldwater temperatures pose a considerable stressor to many fish and few are able to survive water temperatures much below 0-1°C. this is often a major problem in aquaculture in cold climates. Interestingly, some marine teleosts have high levels (10 Ð 25 mg/ml) of serum antifreeze proteins (AFP) or glycoproteins (AFGP) which effectively reduce the freezing temperature by preventing ice-crystal growth. The isolation, characterisation and regulation of these antifreeze proteins particularly of the inter flounder Pleuronectas americanus has been the subject of research for a considerable period in Canada. Consequently, the gene encoding the liver AFP from winter flounder was successfully introduced into the genome of Atlantic salmon where it became integrated into the germ line and then passed onto the off Ð spring F3 where it was expressed specifically in the liver (Hew et al., 1995).The introduction of AFPs to gold fish also increased their cold tolerance, to temperatures at which all the control fish died (12 h at 0o C; Wang et al., 1995). Similarly, injection or oral administration of AFP to juvenile milkfish or tilapia led to an increase in resistance to a 26 to 13°C. drop in temperature (Wu et al., 1998). The development of stocks harbouring this gene would be a major benefit in commercial aquaculture in counties where winter temperatures often border the physiological limits of these species.

The most promising tool for the future of transgenic fish production is undoubtedly in the development of the embryonic stem cell (ESC) technology. There cells are undifferentiated and remain totipotent so they can be manipulated in vitro and subsequently reintroduce into early embryos where they can contribute to the germ line of the host. This would facilitate the genes to be stably introduced or deleted (Melamed et al., 2002). Although significant progress has been made in several laboratories around the world, there are numerous problems to be resolved before the successful commercialisation of the transgenic brood stock for aquaculture. To realise the full potential of the transgenic fish technology in aquaculture, several important scientific break Ð through are required. There include (i) more efficient technologies for mass gene transfer (ii) targeted gene transfer technologies such as embryonic stem cell gene transfer (iii) suitable promoters to direct the expression of transgenes at optimal levels during the desired developmental stages (iv) identified genes of desireable traits for aquaculture and other applications (v) informations on the physiological, nutritional, immunological and environmental factors that maximise the performance of the transgenics of the transgenics and (vi) safety and environmental impacts of transgenic fish.

Chromosome Engineering

Chromosome sex manipulation techniques to induce polyploidy (triploidy and tetraploidy) and uniparental chromosome inheritance (gynogenesis and androgenesis) have been applied extensively in cultured fish species. These techniques are important in the improvement of fish breeding as they provide a rapid approach for gonadal sterilization, sex control improvement of hybrid viability and clonation. Most vertebrates are diploid meaning that they possess two complete chromosome sets in their somatic cells. Polyploidy individuals possess on or more additional chromosome sets, bringing the total to three in triploids, four in tetraploids and so on. Induced triploidy is widely accepted as the most effective method for producing sterile fish for aquaculture and fisheries management. The methods used to induce triploids and other types of chromosome set manipulations in fishes and the applications of these biotechnologies to aquaculture and fisheries management are well described. Tetraploid breeding lines are of potential benefit to aquaculture, by providing a convenient way to produce large numbers of sterile triploid fish through simple interploidy crosses between tetraploids and diploids. Although tetraploidy has been induced in many finfish species, the viability of tetraploids was low in most instances.

In teleosts, technique for inducing sterility include exogenous hormone treatment (Hunter and Donaldson, 1983) and triploidy induction (Thorgaard, 1983). The use of hormone treatements, however could be limited by governmental regulation and a lack of consumer acceptance of hormone treated fish products. Triploidy can be induced by exposing eggs to physical or chemical treatment shortly after fertilization to inhibit extrusion of the second polar body, triploid fish are expected to be sterile because of the failure of homologous chromosomes to synapse correctly during the first meiotic division. Methods of triploidy induction in clued exposing fertilized eggs to temperature shock (hot or cold), hydrostatic pressure shock or chemicals such an colchicines, cytochalasin-B or nitrous oxide.

Triploid can also be produced by crossing teraploids and diploids. Tetraploid induction involves fertilizing eggs with normal sperm and exposing the diploid sygote for physical or chemical treatment to suppress the first mitotic division. Gynogenesis is the process of animal development with exclusive maternal inheritance. The production of gynogenetic individuals is of particular interest to fish breeders because a high level of inbreeding can be induced in single generation.

Gynogenesis may also be used to produce all Đ female populations in species with female homogamety and to reveal the sex determination mechanisms in fish. It is convenient to use all female gynogenetic progenies (instead of normal bisexual progenies) for sex inversion experiments. Methodologies combining use of induced gynogenesis with hormonal sex inversion have been developed for several aquaculture species (Gomelsky et al., 2000). Androgenesis is the process by which would have commercial application in aquaculture. It can also be used in generating homozygous lines of fish and in the recovery of lost genotypes from the crypreserved sperms. Androgenetic individuals have been produced in a few species of cyprinids, cichlids and salmonids.

Biotechnology and Fish Health Management

Disease problem area major constraint for development of aquaculture. Biotechnological tools such as molecular diagnostic methods, use of vaccines and immunostimulants are gaining popularity for improving the disease resistance in fish and shellish species world over for viral diseases, avoidance of the pathogen in very important in this context there is a need to rapid method for detection of the pathogen. Biotechnological tools such as gene probes and polymerase chain reaction (PCR) are showing great potential in this area. Gene probes and PCR based diagnostic methods have developed for a number of pathogens affecting fish and shrimp (karunasagar, 1999). In case of finfish aquaculture, number of vaccine against bacteria and viruses have been developed. Some of these have been conventional vaccines consisting of killed microorgansism but new generation of vaccine consisting of protein subunit vaccine genetically engineered organism and DNA vaccine are currently under development.

In the vertebrate system, immunization against disease is a common strategy. However the immune system of shrimp is rather poorly developed, biotechnological tools are helpful for development of molecule, which can stimulate this immune system of shrimp. Recent studies have shown that the non specific defence system can be stimulated using, microbial product such as lipopolysacharides, peptidoglycans or glucans (itami et al 1998). Among the immunostimulants known to be effective in fish glucan and levamisole enhance phagocytic activities and specific antibody responses (Sakai, 1999).

Cryopreservation of Gametes or Gene Banking

Cryopreservation is a technique, which involve long-term preservation and storage of biological material at a very low temperature usually at -196 C, the temperature of liquid nitrogen. It is based on the principle that very low temperature tranquilize or immobilise the physiological and biochemical activities of cell, thereby making it possible to keep them viable for very long period.

The technology of cryopreservation of fish spermatozoa (milt) has been adopted for animal husbandary. The first success in preserving fish sperm at low temperature was reported by Blaxter (1953) who fertilizes Herring (*Clupea herengus*) eggs with frozen thawed semen. The spermatozoa of almost all cultivable fish species has now been cryopreserved (Lakra 1993). Cryopreservation overcomes problems of male maturing before female, allow selective breeding and stock improvement and enables the conservation (Harvey, 1996) One of the emerging requirement for that can be used by breeders for evolving new strains. Most of the plant varieties that has been produced are based on the gene bank collections. Aquatic gene bank however suffers from the fact that at present it is possible to cryopreserve only the male gametes of finfishes and there in no viable technique for finfish eggs and embryos. However, the recent report on the freezing of shrimp embryos. However, the recent report on the freezing of shrimps embryos by subramoniam and newton (1993) and Diwan and kandaswami (1997) look promising. Therefore, it is essential that gene banking of cultivated and cultivable aquatic species be undertaken expeditiously.

Biotechnological research and development are growing at a very fast rate. The biotechnology has assumed greatest importance in recent years in the development of fisheries, agriculture and human health. The science of biotechnology has endowed us with new tools and tremendous power to create novel genes and genotypes of plants, animals and fish. The application of biotechnology in the fisheries sector is a relatively recent practice. Neverthless, it is a promising area to enhance fish production. The increased application of biotechnological tools can certainly revolutionise our fish farming besides its role in biodiversity conservation. The paper briefly reports the current progress and thrust areas in the transgenesis, chromosome engineering, use of synthetic hormones in fish breeding, biotechnology in health management and gene banking.

Chapter 2

Pain in Fish

The presence of pain in an animal, or another human for that matter, cannot be known for sure, but the possibility that animals experience pain can be inferred through comparative brain physiology and physical and behavioural reactions. Some specialists currently believe that all vertebrates feel pain, and that certain invertebrates, like the octopus, might too. Animal protection advocates have raised concerns about the possible suffering of fish caused by angling. In light of recent research, some countries, like Germany, have banned specific types of fishing.

The idea that animals might not feel pain as human beings feel it traces back to the 17th-century French philosopher, René Descartes, who argued that animals do not experience pain and suffering, because they lack consciousness. Bernard Rollin of Colorado State University, the principal author of two U.S. federal laws regulating pain relief for animals, writes that researchers remained unsure into the 1980s as to whether animals experience pain, and veterinarians trained in the U.S. before 1989 were simply taught to ignore animal pain. In his interactions with scientists and other veterinarians, Rollin was regularly asked to "prove" that animals are conscious, and to provide "scientifically acceptable" grounds for claiming that they feel pain. Carbone writes that the view that animals feel pain differently is now a minority one. Academic reviews of the topic are more equivocal, noting that although the argument that animals have at least simple conscious thoughts and feelings has strong support; some critics continue to question how reliably animal mental states can be determined. Veterinary medicine uses, for actual or potential animal pain, the same analgesics and anesthetics used in humans.

Research Findings

Behaviour: Experiments by William Tavolga provide evidence that fish have pain and fear responses. For instance, in Tavolga's experiments, toadfish grunted when electrically shocked, and over time they came to grunt at the mere sight of an electrode. Additional tests conducted at both the University of Edinburgh and the Roslin Institute, in which bee venom and acetic acid was injected into the lips of rainbow trout, resulted in fish rubbing their lips along the sides and floors of their tanks, which the researchers believe was an effort to relieve themselves of pain. One researcher argues about the definition of pain used in the studies.

In 2003, Scottish scientists at the University of Edinburgh performing research on rainbow trout concluded that fish exhibit behaviours often associated with pain, and the brains of fish fire neurons in the same way human brains do when experiencing pain. Professor James D. Rose of the University of Wyoming criticized the study, claiming it was flawed, mainly since it did not provide proof that fish possess "conscious awareness, particularly a kind of awareness that is meaningfully like ours". Rose argues that, since the fish brain is different from ours, fish are probably not conscious in the manner humans are, and while fish may react in a way similar to the way humans react to pain, the reactions in the case of fish have other causes. Rose had published his own opinion a year earlier arguing that fish cannot feel pain because they lack the appropriate neocortex in the brain. Studies indicating that fish can feel pain were confusing nociception (responding to threatening stimulus) with feeling pain, says Rose. "Pain is predicated on awareness. The key issue is the distinction between nociception and pain. A person who is anaesthetised in an operating theatre will still respond physically to an external stimulus, but he or she will not feel pain." However, animal behaviourist Temple Grandin argues that fish could still have consciousness without a neocortex because "different species can use different brain structures and systems to handle the same functions."

Figure: Goldfish

In a 2009 paper, Janicke Nordgreen from the Norwegian School of Veterinary Science, Joseph Garner from Purdue University, and others, published research which concluded that goldfish do feel pain, and that their reactions to pain are much like those of humans. "There has been an effort by some to argue that a fish's response to a noxious stimulus is merely a reflexive action, but that it didn't really feel pain," Garner said. "We wanted to see if fish responded to potentially painful stimuli in a reflexive way or a more clever way." The fish were divided into two groups, one given morphine and the other saline. They were then subjected to unpleasant temperatures. The fish that were given saline subsequently acted with defensive behaviours, indicating anxiety, wariness and fear, whereas those given morphine did not. Nordgreen said that the behavioural differences they found showed that fish feel both reflexive and cognitive pain. "The experiment shows that fish do not only respond to painful stimuli with reflexes, but change their behaviour also after the event," Nordgreen said. "Together with what we know from experiments carried out by other groups, this indicates that the fish consciously perceive the test situation as painful and switch to behaviours indicative of having been through an aversive experience."

Anatomy

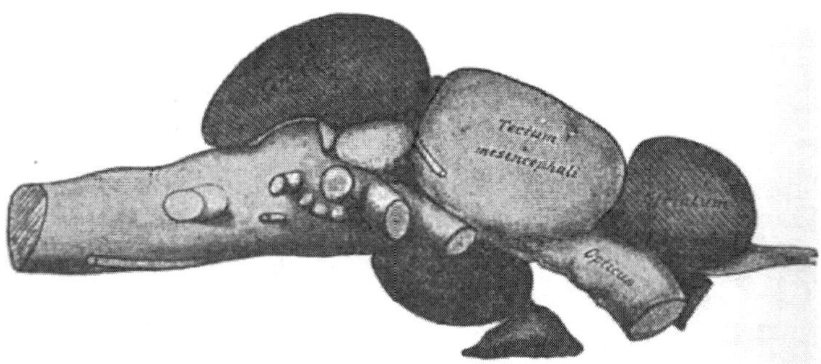

Figure: The brain of a cod

Nociception is the unconscious detection by the nervous system that damage is occurring somewhere. Nociceptors are sensory receptors that respond to potentially damaging stimuli by sending nerve signals to the spinal cord and brain. In 2003 Victoria Braithwaite was able to demonstrate the presence of nociceptors on the face and snout of the trout. The receptors responded to touch, heat and chemical stimulation by sending an electrical signal through the trigeminal nerve to the brain. The Norwegian Research Council is funding a three-year research project, scheduled to end in December 2011, into whether cod can feel pain. The researchers will use fMRI and EEGs to study how the cod brain works. The aim of the study is to identify the parts of the cod brain that activate when cod are exposed to potentially painful stimuli, and how those signals are processed.

Laboratory Fish

Zebrafish, native to the streams of the south eastern Himalayan region, are commonly used as a model organism in studies of vertebrate development and gene function. Zebrafish are used to study development, toxicology and toxicopathology, because the body of a young zebrafish is nearly transparent, providing unique visual access to their internal anatomy. Another extensively used model organism is the medaka, which is much sturdier than the traditional zebrafish. Medakas are easy to rear in the laboratory because of their prolific reproduction rates and short generation times. The short-lived ram cichlid is also used in laboratory studies because of its ease of breeding and predictable pattern of ageing. Sticklebacks have traditionally been used as model organism in the study of fish behaviour.

The extent to which animal research causes pain to laboratory animals is the subject of much debate. Marian Stamp Dawkins defines "suffering" in laboratory animals as the experience of one of "a wide range of extremely unpleasant subjective (mental) states." The United States Department of Agriculture defines a "painful procedure" in an animal study as one that would "reasonably be expected to cause more than slight or momentary pain or distress in a human being to which that procedure was applied."

Artificial Reproduction of Fish

Gonadal Development in Captivity

The female African catfish has a fully developed ovary which contains "ripe" eggs the whole year round, if kept in ponds and once the water temperature remains above 22° C. The eggs of a "ripe" female make up 15-20% of the body weight (i.e. a "ripe" female of 1 kg having about 150-200 g of "ripe" eggs). The oocyte development decreases once the temperature drops below 22°C as can be seen from the results obtained in the Republic of the Congo.

In the dry season (June-July-August) the water temperature drops below 22°C and we see that the ovary usually makes up approximately 5% of the body weight of the female. Artificial reproduction is still possible but the number of eggs obtained is small and the quality of the eggs decreases as can be seen from the decreased hatching percentage.

In general the testis of a male is fully developed at an age of 8 to 12 months once they reach a weight of approximately 200 g. In the Republic of the Congo, sperm could be obtained the whole year round and no effects of temperature on the availability of sperm was found.

In captivity the African catfish does not spawn spontaneously since the environmental factors such as the rise in water level and inundation of shallow areas do not occur on the fish farms. Since the early seventies several techniques have been developed (with or without hormone treatment) for the artificial reproduction of the African catfish.

Induced Propagation without Hormone Treatment

Mature breeders can be reproduced artificially by simulating the events which will occur during the rainy season which trigger off the mating and spawning processes. Ponds of about 400 m^2 in size are filled with 25 cm of water and stocked with six mature females (average weight 300-500 g) and four males (average weight 200-500 g). A few hours later the water level is raised up to a level of 50-60 cm. Spawning will occur at night and the following morning the breeders can be removed.

Theoretically speaking this technique can be termed artificial reproduction since some offspring are produced. However from a practical point of view it is not satisfactory as the number of fingerlings which can be harvested after 6-8 weeks is low (1-2 fingerlings/m^2). This technique of semi-natural reproduction has also been applied successfully using concrete tanks (C. Nugent, *pers. com.*).

Semi-Artificial Reproduction through Hormone Treatment

Hormone Induced Reproduction in Concrete Tanks with a Gravel Substrate

For hormone induced reproduction (semi artificial or artificial) the following hormones are generally used;

- DOCA (Desoxycorticosteroid Acetate), 2.5-5 mg per 100 g of female. A disadvantage of using this hormone is that it is mostly suspended in oil which causes severe ulcers on the injected female.
- HCG (Human Chorionic Gonadotropin), 25 I.U. per 100 g of female. This hormone works well but is expensive.
- Common carp (*Cyprinus carpio*) pituitary gland material, 3-4 mg/kg of female or 1-2 whole pituitaries per female. In

general the common carp pituitary gland material has to be imported from abroad which means that it is not usually accessible to small fish farms.

- Pituitaries of the African catfish (*Clarius gariepinus*). A female catfish will respond once it is injected with a pituitary of a catfish (male or female) of equal size.
- Pituitaries of the Nile tilapia (*Oreochromis niloticus*), 3-4 pituitaries of a Nile tilapia (100-150 g) per female catfish will induce ovulation.
- Pituitaries of Nile perch (*Lates niloticus*). 1-2 pituitaries per female catfish will induce ovulation.

Hormone Induced Reproduction in Ponds

Mature females are injected with hormones in order to provoke the mating and spawning processes and then placed in completely filled ponds at a density of two females and one male per 100 m^2. Spawning usually takes place during the night after the hormone treatment and the breeders then removed the following morning. As for all nursing processes in ponds, the survival in the nursery ponds is a limiting factor for the mass production of fingerlings.

Hormone Induced Reproduction in a Happa Placed in Ponds

A mature female is injected with hormones, so as to provoke ovulation, mating and the spawning processes. After injection the female and a non-injected male are then placed together in a 2-3 m^3 happa (made from mosquito netting with a 0.5 mm mesh size) located within a pond. Spawning usually takes place at night and the breeders then removed the following morning. One advantage of this method is that the eggs are concentrated within the happa where they eventually can be treated against fungal infections and the hatchlings easily collected after yolk sac absorption. It is essential that the happa be covered so as to prevent the breeders from jumping out.

Hormone Induced Reproduction in Concrete Tanks with a Gravel Substrate

This was one of the first techniques developed for the hormone induced reproduction of *C. gariepinus* in Africa (Micha, 1973; van der Waal, 1974).

Mature females are injected in the afternoon with hormonal material (DOCA, HCG or pituitaries) and placed together with a male in a concrete tank. The bottom of the concrete tank is covered with a layer of stony gravel which functions as a substrate for the released eggs. The breeders usually spawn at night after the hormonal material is injected; the released fertilized eggs sticking to the gravel or to any other substrate provided, or on the bottom of the tank if no substrate is available.

The following morning the breeders are then removed from the tank. The fertilized eggs remain in the tank and hatching takes place after 24 hours (25-30°C) and after 3-4 days approximately 3 000-5 000 larvae per female can be collected from the tank. However, the disadvantages of this method are that:

- brood fish often injure each other, which sometimes may result in the death of one of the breeders;
- the number of larvae obtained is relatively small as the ovulation is often only partial; the quantity of eggs released usually being only 5-10% of the total body weight, which is considerably less than the 15-20% which can be usually obtained through stripping.

Artificial Reproduction

Semi-natural or hormone induced reproduction within ponds or concrete tanks as described above can be used on small farms to produce their own larvae and fingerlings. However, this method has not proved to be reliable for the mass production needed for larger fish farms or fingerling distribution centres. Therefore artificial propagation techniques under more controlled conditions, including stripping of eggs, collection of sperm, followed by egg fertilization, have been developed.

Artificial reproduction by induced breeding through hormone treatment followed by artificial fertilization and incubation of fertilized eggs and the subsequent rearing of larvae to fingerling size has several advantages (Woynarowich and Horvath, 1980) including:

- Better rates of fertilization and hatching.
- Protection against enemies and unfavourable environmental conditions.
- Better conditions for growth and survival.

The artificial reproduction of the African catfish, as for all finfishes, consists of a chain of activities which is more or less similar to that which occurs during the course of natural reproduction.

Broodstock Care and Selection of Ripe Breeders

In most cases broodstock selected from nature or bought at a fish farm are kept in earthen ponds at a stocking density of 0.5-1/m^2 and fed regularly with agriculture waste products and sometimes with trash fish. Egg development will take place and about six weeks after a female has spawned it can be used again. In some areas of Africa water temperatures may fall below 22°C during the dry/winter season, which in turn hampers egg development and artificial reproduction. In the Republic of the Congo, this problem was overcome by careful planning i.e. catfish were reproduced artificially in the first month of the dry season and a double number of females were injected in order to guarantee a sufficient number of eggs with the result that catfish fingerlings could be produced 11 months per year. Another method to overcome this problem is to keep the broodstock permanently indoors in a hatchery (de Graaf, 1989, Janssen, 1985a, Richter *et al.*, 1987). A complete breakdown of the natural annual reproductive cycle can be obtained after the broodstock have been kept for one year indoors and reproduction can then be carried out continuously throughout the year. However this method is not recommended by the authors as it depends on the availability of high quality feed and often encounters diseases such as crackhead and retarded growth in the breeders, and an oedemic disease in the developing larvae.

Artificial reproduction starts with the selection of females from broodstock ponds, after which they are transferred to a holding tank within a hatchery. Ideally, broodfish weigh between 300-800 g, with larger fish being difficult to handle and often resulting in substantial egg losses prior to stripping. In general, mature females are selected according to the following criteria:

- well distended, swollen abdomen from which ripe eggs can be obtained by slightly pressing the abdomen toward the genital papilla. Ripe eggs are generally uniform in size and a experienced hatchery operator can see the nucleus as a small dark point in the centre of the egg.
- A swollen, sometimes reddish or rose coloured genital papilla.

Fishes in Space

Odd Behaviour of Fishes in Space

As far as I know, the first time fish behaviour was observed in the zero-gravity environment of outer space was in 1973, when a couple of mummichogs, *Fundulus heteroclitus*, were flown in a plastic bag aquarium aboard Skylab. The crew regularly checked how their charges were doing, and they actually filmed the fish's behaviour on day 3 and day 22 of the mission. On day 3, both fish incessantly "dove" – pitched downward – and thus swam in tight circles, as if stuck to the hands of a clock, a behaviour for which the name "looping" was coined. The frequency of looping decreased steadily on subsequent days, until it eventually disappeared. When the fish were filmed again on day 22, they both swam normally with their backs turned towards the cabin's light source (this is a behaviour known as the "dorsal light response" – more on that later). However, episodes of looping could still be triggered by gently shaking the bag aquarium. Fifty eggs at an advanced stage of development had also been taken on board, and 48 of them hatched during the flight.

In a follow-up study, 21-day-old mummichogs were flown on Skylab again, and this time astronauts (on day 9) reported a lack of dorsal light response.2 Other work with carp flown on the space shuttle Endeavour in September 1992 showed a disruption of the dorsal light response for the first 3 days but a gradual recovery thereafter. The dorsal light response was already well known from earthbound observations.

Two mechanisms allow a fish to figure out which way is down (or up) on earth. In the first one, minuscule corpuscles in the inner ear are pulled down by gravity until they set off sensory cells. Depending on the direction of gravity, different cells are stimulated, enabling the fish to know which way is down. This is called the "vestibular righting response", and fishes share this mechanism with land vertebrates, including humans. The second mechanism is simple: the direction where light comes from is interpreted as "up". This is reflected by the tendency of fish to turn their backs towards a light (hence the moniker "dorsal light response"). For fish, light is a good directional cue because in an aquatic environment light usually comes from above and only from above. In terrestrial habitats the ground can reflect light back up, but in any moderately deep body of water no light ever comes from below.

One can demonstrate the dorsal light response by placing a fish in a vertical tube so narrow that the fish has no choice but to take a head-down posture. Then a single light is turned on, the lefthand side for example. If the dorsal light response is well established in that species and in that individual (the phenomenon, though widespread, is not universal), the fish will swivel inside the tube until its back is turned to the light.

Another way to demonstrate the dorsal light response is to look head on at a fish in an ordinary aquarium, again with the only light coming horizontally from the side. Rather than staying perfectly vertical, the fish will slant its back slightly towards the light. The degree of slant can be taken as a measure of the relative importance of light versus gravity as a cue for the fish to determine which way is up. The more slant there is, the more important light is relative to gravity. If gravity is paramount, the body will remain perfectly vertical. If light is paramount, the fish will swim "on its side", with its back exactly towards the light. In that respect, it is worth noting that in Skylab, where there was no gravity, all fish turned their back completely to the light (except, in some cases, during the first few days of the mission). Light was the only cue they had to figure out which way was "up". Looping behaviour was also known before 1975, though not really because of earthbound observations. Instead, it had been witnessed in goldfish taken for a ride on parabolic flights in 1969 and 1972. A parabolic flight is achieved when a plane climbs at a relatively steep angle to a high altitude and then briefly levels off before diving down. The manoeuvre (which, if it could be seen from the side, would describe the general shape of a parabola) creates an upward centrifugal force that completely counteracts gravity (the counteracting is made easy by the fact that gravity at high altitude is weaker). The zero-gravity phase lasts for less than a minute.

All goldfish taken on such flights had looped without fail during the zero-gravityphase. Some fish had also performed spinning movements, like corkscrews. In the same way that a transfer from normal gravity to zero gravity induces looping, raising a fish in conditions of higher than normal gravity (this is done by putting an aquarium in a huge centrifuge and letting it turn for weeks on end) induces looping once the fish is brought back to normal conditions. There is no convincing explanation for why fishes loop or spin.

Motion Sickness in Fishes

Many astronauts become motion sick during the first 2-3 days of a space mission. On earth, motion sickness consists of a malaise (often leading to vomiting) felt when the body is suddenly accelerated or decelerated or when it changes its direction of movement, especially when information coming from the eyes suggests to the brain that no such motion should take place. Sea sickness is a typical example. In the case of astronauts, the symptoms are the same, but the cause is different: the problem arise because of movements performed in weightlessness. In this case the illness is called "space motion sickness". Astronauts eventually adjust and get better after a few days. However, after landing on earth they often go through another bout of sensory-motor disorders, again for a few days. Fish have not been reported to vomit in space or during parabolic flights. However, the occurrence of looping – a quantifiable behaviour easily witnessed – follows a similar timeline to that of space motion sickness. Therefore it is thought that fish could serve as an animal model to study space motion sickness and possibly find ways to alleviate it. Note in passing that people – back on earth – who regularly move fish in transport tanks do sometimes see their fish vomiting while in transit, especially when the tanks are roughly shaken. This seems to be a case of motion sickness. Fish vomiting in transit form the basis of claims to the effect that even fish can get seasick. A more accurate statement would be that even fish can get motion sick.

The First Vertebrate Mating in Space

In the 1990s, a team of Japanese scientists headed by Kenichi Ijiri explored the possibility of sending more fish in space, this time aboard the space shuttle Columbia. The idea was to see if fish could be induced to mate successfully in the absence of gravity (and, by the same token, to provide the first example of a successful vertebrate mating in space). The fish species they chose was the medaka (*Oryzias latipes*), a tough, prolific breeder and a very popular pet fish in Japan. As a preliminary step, the scientists observed the behaviour of medaka during parabolic flights. What they saw was some good old looping. This was not surprising (up to then all fish species had looped in zero or near-zero gravity) but still it was disheartening because it is hard to imagine two fish courting and mating successfully while looping. Even if the medaka were to settle down after a few days, as the mummichogs had done aboard Skylab, they might still not mate because of the general exhaustion and lack of eating resulting from all the looping. But the scientists persevered: they took a great number of medaka on parabolic flights until they found some that did not loop. Then back in the lab they bred those few individuals to create a strain of non-loopers from which they could select the future medaka-astronauts. Interestingly, these fish refrained from looping only when there was light. If kept in the dark during parabolic flights, all medaka looped (the scientists observed the fishunder infrared light and with infrared goggles). Ijiri and his team posited that nonloopers were particularly predisposed towards using light as a cue for maintaining position. Indeed, when submitted to visual tests, the nonloopers scored particularly well. (One such test consists of placing a fish in a circular tank with rotating walls.

The walls are painted with vertical stripes. When the walls rotate, the fish have a tendency to follow the stripes and thus swim around the tank – this is the so-called "optomotor response", often explained with the argument that trying to stay at a constant distance from a landmark may be a way to maintain position in a current. Fish with good vision keep swimming around the tank even when the walls rotate very quickly, whereas fish with poorer vision soon see the rotating stripes as a blur and stop moving.)

When medaka are in the mood, they can mate and produce eggs every day. The Japanese researchers selected two non-looping males and two non-looping females who were particularly assiduous at breeding. These were placed in a special enclosed aquarium that was loaded on board the space shuttle Columbia some 30 hours before its launch in July 1994. Lift-off took place without a hitch and already 24 h into the mission a few eggs could be seen inside the aquarium (the aquarium had been built so that a current would sweep any free-floating eggs into a small compartment where a mesh protected the eggs against any cannibalistic attack by the adults). On the third day, a male and a female were caught on video in the typical medaka mating posture, the male clasping the female with his fins. The scene was repeated many times during the whole mission, and eggs were steadily produced. On the 12th day of the 15-day mission, the first egg hatched normally. By the time the shuttle landed, the aquarium contained 11 fry and 27 embryonated eggs. All of these eggs hatched successfully within three days of the landing. Interestingly, for some time after the landing the four adult medaka looked awkward and seemed to have trouble swimming. It took them three days before they returned to normal. All fry, however, swam without any problem. These space-born fry grew up normally and went on mating with one another on earth, and their successive generations have been distributed to elementary schools and school children throughout Japan.

How do Fishes React to Total Solar Eclipses?

People who are interested in the time sense of animals are naturally curious about the reactions of animals to total eclipses of the sun. For the few minutes of totality, when the whole sun disk is obscured by the passing moon, light levels at the surface of the earth are as low as they would be during twilight. If this happens in the middle of the day, a diurnal animal that is out and about will be confronted with contradictory information: the sudden dimming of the light will suggest dusk and tell the animals that it is time to go to bed, but the animal's internal time sense will insist that this is the wrong time of day for sleep.

Watching the reaction of diurnal animals during a solar eclipse can tell us whether the move to a sleeping site is triggered more by ambient light levels or by an internal clock. Published reports indicate that during an eclipse, mammals either keep on going about their normal business (for example, zebras, lions, baboons, ground squirrels, dairy cattle) or appear to be nervous (for example, chimpanzees, horses, hippos). They do not seem to prepare for sleep in any way. We therefore conclude that internal clocks seem to be strongly relied upon. Birds, however, often behave oddly, crowding together and becoming quiet. Some birds have been seen flying to their roost, a presleep behaviour. One observer in Africa reported that waders and herons showed no reaction to a total solar eclipse, but that egrets, oxpeckers, ibis, trumpeter hornbills and geese set off for their roosts.

Meanwhile, nocturnal owls and frogs started calling, as if night was starting. Reptiles and insects can also adopt night time behaviour during totality. During a 1991 total eclipse in Mexico, orb weaving spiders were witnessed taking down their webs, as they do every night, and zebra-tailed lizards also seemed to turn in for the "night". This diversity of reactions does not help us settle the question of whether pre-sleep behaviour is triggered more by light cues or by internal clocks. If all animals remained impassive or just got nervous, as if they knew something was wrong, we could say "Aha! Internal clocks have the upper hand!" On the other hand if all animals went to roost, we'd say "Aha! They rely on light cues only". As it is, it seems that both factors are involved to a certain degree.

Now What About Fishes?

I have found only two scientific papers on the behaviour of fishes during solar eclipses. In the first one 5, three freshwater species of air breathers (the cuchia *Amphipnous cuchia*, the climbing perch *Anabas testudineus*, and the banded gourami *Colisa fasciata*) were observed in captivity during a 1980 solar eclipse in India. Allthree species almost stopped gulping air, became sluggish, and settled to the bottom. *A. cuchia* and *A. testudineus* also changed colour and turned blackish. All this is typical night time behaviour. In the second study, various reef fishes were observed in the wild during a 1998 total solar eclipse at Pinta Island, in the Galapagos. As the eclipse progressed, the fish descended from the water column and retreated to crevices within the reef, where they remained hidden for the duration of totality. This is typical twilight and night time behaviour. The fish did not appear to go to sleep however, for they resumed their normal activities within three minutes of the end of totality. The only diurnal species that did not exhibit night time behaviour during totality was the surgeonfish *Prionurus laticlavius*, which stopped grazing and formed schools that swam erratically with apparent alarm. Altogether, the latest case notwithstanding, these observations suggest that light intensity is fairly important in eliciting twilight and night time behaviour.

It is possible to study the response of fish to a midday dimming of the lights in the laboratory. But first we need to identify a behaviour that is typical of dusk or night time, and easy to measure. One such behaviour is fry-retrieving in cichlids. Many species of cichlids keep watch over their brood of free-swimming fry. When night approaches the parents excavate a pit in the gravel of their aquarium and then they seek their fry, catch two or three at a time in their mouth, return to the pit, and unceremoniously spit the fry into the pit.

They keep doing this until all the fry are retrieved. Thus when night finally arrives the fry are gathered in one place so that the parents can easily guard them. Usually parents start retrieving just before nightfall, and if they are not done by the time all lights go off they continue retrieving in complete darkness. (How they manage to find their young and return to the pit in the dark is an interesting question, as yet unanswered. It could be addressed with the help of infra-red lighting and infra-red goggles to observe the retrieving fish). In one study of this behaviour, I installed parental convict cichlids in tanks under a regime of 12 h of lights-on and 12 h of lights-off. When I dimmed the lights 15 minutes before the normal time of lights-off, my cichlids busily retrieved their fry.

But when I dimmed the lights in the middle of the day, or in the middle of the afternoon, they did not retrieve. It looked as if their internal clock told them that a midday dimming could not be twilight. However, there were limits: when I completely turned off the lights in the middle the day, then the parents started to retrieve their fry. Fry-retrieving seems to be partly influenced by light levels, and partly by an internal clock that tells the fish when it should be done. And with this we have reached the same conclusion as with the observations of other animals during eclipses. It is not black and white, but it's the best we can do for now.

Sleep in Fishes

Do fishes sleep? This is not an easy question. It depends on what you mean by "sleep". Most sleep researchers, being focused on mammals, recognise sleep when they see distinctive patterns of electrical activity of the brain. Fish, however, lack the complex brain structures from where these patterns originate. Another criterion to define sleep, used mainly in the field or in situations where electroencephalographs are not available, is prolonged eye closure. But fish do not have eyelids! Fortunately, other criteria exist that can be expanded throughout a vast array of animal classes, including fish. These are (1) a state of prolonged inactivity, (2) a typical resting posture, usually in a typical shelter, (3) 24-h rhythmicity, and (4) high arousal thresholds.

Thus, most people would agree that an animal is, for all practical purposes, asleep when it remains almost completely immobile (criterion # 1), in almost always the same posture and same location (# 2), at the same time every day (# 3), and in a state of decreased sensitivity to disturbances (# 4). A few scientists have applied these criteria to fishes and taken the question to the laboratory. Colin Shapiro reported that his Mozambique tilapias, *Oreochromis mossambicus*, used to rest on the bottom of their tank at night, with a lower respiratory rate and no detectable eye movement, and that they did not respond as readily as during the day to stimuli such as an electrical current or the delivery of food. Similarly, at the Academy of Sciences of the USSR in Leningrad, I.G. Karmanova and his team witnessed brown bullheads *Ameiurus nebulosus* in a typical sleep posture, with the tail lying flat on the ground, the other fins stretched out, and the body inclined to one side at an angle of 10-30o to the vertical. Cardiac and respiratory rhythms were much slower than normal and the fish were less sensitive to sound and tactile stimulation.

The researchers touched the bullheads with glass rods and the catfish did not move. This certainly fulfils the criterion of reduced sensitivity. In fact, when it comes to lack of sensitivity during sleep, there might be even better examples: several researchers have said that the Spanish hogfish *Bodianus rufus*, the bluehead *Thalassoma bifasciatum*, the wrasse *Irideo bivittata* (which is probably the slippery dick *Halichoeres bivittatus*), the cunner *Tautogolabrus adspersus*, and even requiem sharks could be picked up by hand at night and even sometimes lifted all the way up to the surface without evoking a response.

In studies of this kind, one has to be mindful of the lighting that is used during the nocturnal observation. There is a phenomenon called "light shock" which happens in some fishes – not all!– when they are exposed to a sudden bright light at night. In response to the light, these fishes sink to the bottom and remain motionless for several minutes. This may give the illusion of sleep when in fact it is only a shortlived reaction to a temporally abnormal stimulus. For experiments on sleep, it is preferable to use very dim illumination constantly throughout the night. Another solution is to resort to near-infrared technology. I have obtained infrared goggles from an Army Surplus store and used them in conjunction with a powerful flashlight fitted with an infrared filter to unobtrusively observe captive fishes in complete darkness.

I concentrated on members of the family Cichlidae, as they are known to be blind to infrared light. What I found was that these fish were indeed quiescent at night. Adult oscars were sluggish and unresponsive, resting on the bottom with their eyes turned downwards. Severum cichlids *Heros severus* were also quiescent, either near the bottom or higher up in the water column. Rainbow cichlids *Herotilapia multispinosa* and convict cichlids *Archocentrus nigrofasciatus* were often immobile as well, with their pelvic fins lowered all the way down, but occasionally they set off on slow promenades with their pelvic fins brushing against the gravel bottom. Yet another solution to the problem of artificial light at night is to switch criteria and to concentrate on the use of typical resting places (most sleeping fishes will not leave their shelter just because they are suddenly lit by a bright light – if anything, they will dig in). Based on this criterion alone, many fishes do seem to sleep. Divers in the field have often seen fishes "settling down for the night" in holes and crevices, underneath ledges, amidst vegetation or corals, inside sponges, or buried in sand. Such shelters were found to harbour fish throughout the night. Some extra protection can be afforded by special secretions, such as the mucous envelope that is produced by several species of wrasse and parrotfish, either around the fish themselves or at the opening of their shelter. Spotted moray eels prey more heavily on sleeping parrotfish that do not secrete mucous envelopes as compared to those that do. Presumably the mucous cocoon masks the odour of the fish and makes it less detectable.

Another sleep criterion is the existence of a period of inactivity that alternates with a period of activity on a 24-h basis. Here, direct observation is tough: either you stay up at all hours of the day and night, many days in a row, just to watch fishes sleep, or, if you filmed the fish's behaviour, you spend hours upon hours analysing videotapes. Who really wants to do that? Instead, researchers resort to automated activity-recording systems.

The most common of these systems involves infrared light beams criss-crossing the aquarium. When a fish swims into the path of a beam, it cuts the electrical circuit linked by that beam, and such interruptions can be recorded by computer, day and night continuously. Another method consists of setting up an ultrasonic wave throughout the aquarium. Large movements on the part of a fish stir the water and alter the wave. Given that the wave received at one end of the aquarium can be electronically compared to the one generated at the other end, and that any difference can be made to create a voltage pulse recorded automatically on computer, the large movements of the fish can be continuously monitored. There is also a third method, reserved for those who study electric fishes, and it is simply to record the occurrence of electrical discharges in the aquarium. This is based on the fact that some electric fishes discharge at low rates when they are inactive, and high rates otherwise.

Other methods can be used in the field. Researchers can set traps and nets, and check them every few hours or so, day and night. Sleeping fishes are not supposed to swim into traps, but they often do so when they are active, and therefore a 24-h pattern of captures can be interpreted as a 24-h pattern of activity. One can also use fishing implements that catch both active and inactive individuals (a seine net, for example), and then dissect a subsample of these fish to look at their stomach contents, the idea being that sleeping fishes do not feed and their stomach should be empty. A 24-h pattern of stomach fullness can therefore be interpreted as a 24-h pattern of activity.

Now, for the sleep researcher, all of these lab and field methods share one disadvantage: they record only one criterion of sleep – behavioural quiescence. But for yielding continuous information, they are fairly reliable, and accordingly they have been used quite extensively. The scientific literature is replete with studies of 24-h activity patterns. As expected, most fishes are found to be either diurnal or nocturnal, although inactivity during the rest period is seldom absolute. In diurnal fishes for example, we can still find a few captures, a few food items in the stomach, or a few beam interruptions at night. Obviously, we are not talking about sound sleep for every member of the population here. And to complicate things further, a fair number of species seem to be crepuscular, specialised primarily for the twilight conditions of dawn and dusk. Such crepuscular species are often predatory, taking advantage of the increased traffic that occurs at dusk when diurnal fishes hurry to their resting places while nocturnal fishes emerge out of theirs, or vice-versa at dawn. Indeed, it would be a mistake to believe that all fishes can sleep, or that sleep in fishes is as absolute as it is in mammals. So now we come to the matter of exceptions, which are numerous (they may not deserve to be called exceptions) and can be summarised as follows: (1) some fishes apparently never sleep, (2) some fishes may not sleep in early life, although they do so as adults, (3) some fishes do not sleep during migration, (4) some fishes do not sleep during the spawning season, (5) some fishes do not sleep while caring for their young, and (6) some fishes are capable of switching from night-rest to day-rest. Let us review the evidence for these statements.

Complete Lack of Sleep

The observation that a fish never seems to sleep often tends to be dismissed as an artefact of captive conditions and the stress such conditions impose. Either that or it is not reported in the literature because sleep researchers are in the business of studying concrete aspects of sleep and not the lack of it. Yet, the number of fish species that never sleep may be greater than we think. There are, in fact, a few mentions of non-sleeping species here and there in the scientific literature. For example, Bori Olla and Anne Studholme have reported that bluefish *Pomatomus saltatrix* and Atlantic mackerel *Scomber scombrus* keep on swimming day and night in large seawater aquaria, and although these fish swim more slowly at night, they remain responsive to disturbances or the introduction of food 24 hours a day. Could it be that some pelagic (open water) species need to swim all the time, either to breathe ("ram ventilation" of their gills) or to keep afloat, or to prevent being swept away by the current, and this is incompatible with sleep? Possibly. Certainly this argument has been put forward not only for mackerels but also for tunas, bonitos, and some sharks. 16 Some species do not need to swim all the time but they remain constantly alert nonetheless. For example many coral reef species are active at night, feeding, and during the day they form schools that are stationary but still responsive to predators. Another case is that of the nocturnal California horn shark, *Heterodontus franscisci*. At night it swims around and feeds, but during the day it hides in caves. Despite being inactive in such a shelter, it still responds very quickly when disturbed by divers. (Astute readers will notice that these sharks fulfil at least two criteria of sleep – use of a typical shelter on a 24-h basis – but not another – decreased sensitivity to disturbances. Truly, it is not easy to define sleep in fishes.)

Lack of Sleep During the Juvenile Stage

Studying the sleeping habits of a fish throughout its whole life has not been done very often – boredom takes hold of you after a while! The only example that I am aware of is a study by Colin Shapiro and his colleagues on the Mozambique tilapia. After a few long days and sleepless nights spent in constant observation, these researchers came to the conclusion that juvenile tilapia do not show any sign of sleep, at any time. In fact it took these young fish 22 weeks to develop the adult sleep patterns. As of yet nobody knows if such lack of sleep in the early stages of life is common place among fishes, and no adaptive explanation for this delayed development has been put forward.

Lack of Sleep During Migration

Many birds do not sleep during migration, and the same may be true of fishes. Olla and Studholme have looked at the activity patterns of a demersal (bottom-dwelling) wrasse, the tautog *Tautoga onitis*. In summer, in the lab, these fish were active only by day. However, when photoperiod and temperature were adjusted to simulate fall conditions, a time when adults normally migrate offshore to overwinter, Olla and Studholme's fish became active day and night, if not more so at night. Such continuous activity did not leave much room for sleep. Interestingly, this increase in nocturnal activity was observed in adults only, not in juveniles. This is consistent with the fact that juveniles do not migrate in the fall, preferring instead to stay inshore to overwinter. As an additional example, there is the case of lake chubs *Couesius plumbeus* living in streams. These fish are caught in minnow traps only at dawn, dusk, or during the day in summer, an observation that classifies them as diurnal/crepuscular. However, at the time of in-stream (spring) or out-stream (fall) migrations, counting fences set up across the waterway intercept these fish mostly at night. Fishes moving within streams must go over many shallow areas where they are probably more visible and more susceptible to predation by kingfishers, and it may therefore be safer to migrate at night, even if it means going without sleep.

Lack of Sleep During the Spawning Season

It seems that mating in fishes cannot be restricted to just one half of the day. 24-h patterns of activity often break down during the spawning season in fishes. Spawners become active day and night. Examples include the tautog mentioned above, as well Disadvantage: they record only one criterion of as yellow perch, minnows, various centrarchids, bullheads, river lampreys, and others. I have observed diurnal convict cichlids showing signs of being ready to spawn at the end of the day, and turning off the lights for the night did not frustrate them. They kept on digging pits and shoving gravel away in complete darkness (even the pilot lights of the heaters had been disabled – my infrared goggles came in handy here). In complete darkness the fish nudged and skimmed the spawning substrate in apparent cleaning movements, and they even laid eggs! The broods of such nightspawners tended to be more scattered on the substrate than those of day-spawners, but they did not comprise more unfertilised eggs.

Lack of Sleep During the Parental Phase

Human parents of newborn babies know how hard it is to have to wake up at all hours of the night. They would envy fishes, who seem to be able to forego sleep and stay active 24 h a day when they have to care for their eggs. Some fish species stay with their brood of eggs and fan them, a behaviour that provides a steady supply of oxygenated water to the eggs. Because eggs consume oxygen 24 hours a day, we would expect parents to fan all the time, and that is indeed what they do. Using my infrared equipment once again, I have witnessed convict cichlid females fanning their eggs day and night. In fact, they fanned three times as much at night as during the day – maybe they got interrupted by visual stimuli during the day, or maybe nocturnal fanning is a hard-wired adaptation to lower oxygen availability caused by the absence of oxygen-producing photosynthesis by aquatic plants at night. With a different piece of equipment (a light-magnifying starlight scope), I have also documented high levels of nocturnal fanning by threespine sticklebacks in the wild. And there have been other reports of continuous (day and night) parental care in various damselfishes, in smallmouth and largemouth bass, and in the brown bullhead.

Day/Night Sleep Switches

A notable difference between fish sleep and human sleep is the fact that for many fishes, sleep (or inactivity) does not seem to be tied to a particular phase of the 24-h day. There are a number of documented switches from nocturnalism to diurnalism, or vice-versa, in fishes. Sometimes, part of a fish population is diurnal and part is nocturnal. Even the same individual can be diurnal for a length of time, and then become nocturnal at some point. Some of these switches can occur according to the seasons, particularly up north. In the 1960s and 70s, Karl Müller and Lars-Ove Eriksson, from the University of Umea in Sweden, worked at a field station located at 66o N – above the Arctic Circle – and found that captive burbot, sculpins and brown bullheads, although nocturnal in summer, became diurnal under the short photoperiod of the Arctic winter.

This switch happened only at high latitudes. When tested at field stations below 60° N, the fish remained nocturnal throughout the year. Diurnal species, for their part, did not switch (the researchers looked at European minnows, yellow perch, and various diurnal salmonids). 28 Interestingly, some Arctic voles have also been reported to switch from nocturnalism to diurnalism in winter. The adaptive significance of this phenomenon, however, remains unknown.

In salmonids, temperature may affect the phasing of activity. Neil Fraser and a number of Scottish and Norwegian co-workers have manipulated the activity phase of juvenile Atlantic salmon simply by chilling the water. They started by using infrared light to spy on the behaviour of captive juveniles (parr) at night. These fish could either stay underneath a shelter or come out and look for food. After artificially lowering the water temperature from 18 oC to 2 oC in summer, Fraser and his team saw that the salmon left their shelter less and less often during the day, but still very readily at night. Essentially the fish became proportionally more nocturnal as water temperature decreased. This jibes well with a number of field observations to the effect that young salmon in the wild are less often seen during the day as temperatures plummet in the fall, or that they remain exclusively nocturnal all year long in cold streams. The scientists surmised that cold salmon react less quickly to predators because of their lower metabolism, and so when the mercury takes a dive they tend to become proportionally more nocturnal to avoid the diurnal warm-blooded birds that prey upon them. When it is warmer and they feel better at eluding predators, the salmon become more diurnal to take advantage of the better lighting conditions to catch food.

This interplay between food availability and predation risk can be used to tentatively explain other types of day/night switches. For example, white suckers *Catostomus commersonii* are known to be diurnal when living in a shoal but nocturnal when living alone. They, as well as many other freshwater fishes, also tend to be diurnal as juveniles but nocturnal as adults. What could explain these differences?

Well, perhaps smaller (that is, less visible) and shoaling fish are less vulnerable to predatory birds such as kingfishers and can therefore afford to look for food during the day. Or maybe the smaller food items on which juveniles feed are more available during the day. Such hypotheses still remain to be tested.

There is no doubt that the temporal availability of food can affect activity patterns. Richard Spieler and his collaborators have spent a fair amount of time studying fish rhythms in the back rooms of the Milwaukee Public Museum. In one experiment, Spieler and Teresa Noeske fed some goldfish only at night, others only during the day, and yet others only at dawn or dusk. Then, after a few weeks, the food was withheld for 10 days.

During that period of fasting, each fish turned out to be active only around the hour at which they used to be fed, be it during the day, the night, or at twilight. This could not be a direct response to food because food was not delivered any more. Instead it seems that the fish had learned to adopt a daily phasing of activity that maximised their foraging success. Diurnal, nocturnal, crepuscular, they did not care, as long as they could get food. (A similar manipulation of activity phasing has been attempted with a nocturnal mammal, the weasel. Food was made much easier to obtain during the day, but the weasel remained stubbornly nocturnal, showing less plasticity than fishes.) In humans, sleep is preferably done at night because our senses (vision, especially) are adapted to day-living. But many fish species seem to get by in complete darkness just as well as in broad daylight. For these animals, being strictly diurnal or nocturnal is not an obligation. Accordingly, they can easily switch between being diurnal and being noctunal. Food and predators are probably the main determinants of their preferred phase of activity, and therefore their phase of sleep (if they sleep at all).

Last Word

So, what is the answer to the question "do fishes sleep"? As unsatisfactory as it is, the answer seems to be: it depends! It depends on which and how many criteria you are willing to use to define sleep. And it depends on the species, the age of the fish, its reproductive status, the time of year, the water temperature, the presence of predators, the availability of food, and who knows what other factors that still remain to be investigated. The last word is probably that, although signs of sleep are present in fishes, we should not equate it with the kind of sleep that we, humans, are used to experiencing.

Oxygen and Fish Behaviour

One of the basic principles of good aquarium maintenance is to make sure that the water is well aerated, so that enough dissolved oxygen is available for fish respiration. This oxygen can be produced by the photosynthesis of aquatic plants and algae, or it can simply diffuse from air to water. In most environments it is diffusion from air that represents the main source of dissolved oxygen. That's certainly the case for most aquaria. (In that respect, one should know that filtration systems are useful not only for cleaning the water but also for stirring it, bringing bottom water to the surface where it can take up oxygen from the air before being pushed back to lower levels where the fish can enjoy it. Air bubblers perform the same function and, as an additional benefit, they increase the area of the air-water interface.) In contrast to the ideal conditions that normally prevail in the home aquarium, wild fishes are sometimes faced with situations where little oxygen is present in the water. In places such as swamps and pools, the absence of wind and water currents prevents the mixing of the oxygen-rich upper layer with the rest of the water column. 1 In other places, an overabundance of aquatic animals or decomposing matter depletes oxygen because of increased consumption. These conditions are not uncommon, and therefore it is not surprising to learn that fishes have evolved special behaviours to cope with low oxygen levels. Good aquarists should have no cause to ever witness these behaviours, but biologists, being the curious lot that they are, have sought or created hypoxic (low oxygen) conditions and catalogued the reactions of fishes to them. Here is what they found.

Increased Ventilation

A universal reaction to low oxygen is increased ventilation of the gills. In the same way that people start to breathe faster in order to bring more air to their lungs, fishes increase the rate of water flow through their gills. A convenient measure is the rate at which gill covers open and close under different levels of dissolved oxygen. Thus three-spined sticklebacks have been reported to increase their rate of gill cover movements from 95 to 165 per minute after the oxygen content of their water was experimentally cut down to half of its normal level. One could also hypothesize that ram-ventilating sharks (fishes that bring oxygenated water to their gills by swimming with their mouths open) would increase their swimming speed and would open their mouths wider when oxygen is scarce. This is indeed what has been observed. In the lab, researchers normally reduce oxygen levels by blowing bubbles of a physiologically neutral gas – usually nitrogen – into the water; nitrogen then becomes much more abundant than oxygen and displaces it as a dissolved gas. One has to be mindful of confounding variables in experiments of this kind: disease, stress and strenuous activity can also make fishes – like people – breathe faster. To eliminate the potential effect of these variables, it is important to work with animals that are consistently healthy, calm and not fidgety. Moreover, fishes are ectotherms (coldblooded), which means that warm temperatures can raise their metabolism and consequently their breathing rate also. Comparisons between oxygen levels must therefore be done at similar water temperatures.

Decreased Activity

After encountering an area of low oxygen content, fish often start to swim rapidly and to zig-zag all over the place. This is probably an adaptive response to escape the danger zone. However, if hypoxic conditions persist, most fish will greatly curtail their general activity. The advantage of inactivity in the face of hypoxia is simple: less muscular work means less need for oxygen. As part of his doctoral research at Laval University in Quebec City, Fred Whoriskey observed sticklebacks breeding in tide pools along the St-Lawrence Estuary, and he reported that during periods when dissolved oxygen levels dropped, the fish became very sluggish, even failing to show normal fright responses when an object flew over the pool (the object could be a bird or, if nature did not oblige, a frisbee thrown by the resourceful researcher). In the same vein but this time in a lab at McGill University in Montreal, Don Kramer and his students observed that guppies swim less, eat less, and court less when oxygen levels are lower than normal. Inactivity can even reach extremes, as in the case of the South American cichlid *Biotodoma cupido* (the greenstreaked eartheater) which appears to fall "asleep" when oxygen is scarce.

Aquatic Surface Respiration

Another common response to low oxygen is aquatic surface respiration. In this behaviour, fish stay just below the surface, put their snout at the air-water interface, and breathe in the film of water that is in direct contact with the air. This thin layer of water is comparatively rich in oxygen. Once I have seen this behaviour done by wild sticklebacks in a tide pool. It was a very warm summer night without wind. The water in the pool did not hold its normal load of dissolved oxygen because it was warm, the absence of wind prevented mixing, and aquatic plants could not photosynthesise in the dark and therefore did not produce any oxygen. In the lab, Don Kramer and Martha McClure studied 24 species of tropical fish common to the pet trade, from tetras to barbs to cichlids. They found that all of these fishes performed aquatic surface respiration when oxygen was deficient. Typically, the fishes waited until a very low threshold of oxygen concentration was reached – less than 2 ppm9, or about 25% of well aerated freshwater at room temperature – before starting surface respiration. This reluctance to breathe near the surface is easy to understand when we consider that, in nature, many predators of fishes are terrestrial or aerial animals that attack from above, and therefore being close to the surface is risky.

Air-Breathing

Some fishes can switch from water-breathing to air-breathing when there is not enough oxygen in the water. These fishes break the surface and trap air inside their mouth. Then they use their mouth, pharynx, oesophagus, swimbladder or even in some cases their stomach like we use lungs, that is, as a site where oxygen can diffuse into the blood. Examples of such fishes include gouramis, goldfish, loaches, catfishes, the algae-consuming suckermouth catfish so common in home aquaria, and the lungfishes. Air breathing allows many fishes to lead an amphibious lifestyle, or to temporarily move out of the water if they are bothered by predators, competitors, or the threat that their water world will disappear because of a drought. Among the most amphibious of these fishes are the mudskippers, which spend most of their active lives in the air on intertidal mudflats where they feed and mate and bicker with their territorial neighbours. They are skittish creatures though, and at the slightest hint of danger they will retreat into water-filled burrows dug in the mud. At least two species – the giant mudskipper *Periophthalmodon schlosseri* and the walking goby *Scartelaos histophorus* – exhibit the interesting behaviour of storing air inside their burrow. They take air into their mouth, dive into their waterfilled burrow, and release the air into special chambers. If researchers experimentally extract air from the special chambers, the fish diligently replenish it. The significance of this behaviour stems from the facts that at high tide, when water covers the mudflats, the fish stay in their burrow to avoid predators, and water inside the confined burrow is often poorly oxygenated. At such times these air-breathing fishes can tap into the air reserve of their special chambers. There are several physiological advantages to using air as a source of oxygen. Per unit volume, air contains about 30 times more oxygen than water (the exact value depends on water temperature). Moreover, air is 1000 times less heavy and 50 times less viscous than water, and therefore easier to move through the respiratory system.

To grasp the difficulty of breathing in water rather than air we need only consider the fact that to survive, a fish of 100 g typically must move 30 to 65 g of water per minute through its gills – that is 1/3 to 2/3 of its own body weight, a substantial exercise! Only the incredibly large surface area of the gills can compensate for this disadvantage. But breathing in air also carries a major disadvantage: air cannot support the gills. Gills are thin to facilitate gas exchange and to maximise their number and surface area. Because of their thinness, the gills usually are not selfsupporting. For a strictly water-breathing fish this is not a problem because water provides support for the gills. However, if such a fish finds itself out of water, the gills are not supported anymore. The collapse of gills and the limited surface area of other possible breathing organs are the reasons why most fishes ultimately choke outside water, despite the abundance of oxygen in air. Only air-breathing fishes can manage, and among their anatomical adaptations are thicker, self-supporting (but, consequently, less numerous) gills, as well as modified swimbladders, mouths, gill chambers, guts, or skin which have more blood vessels and are more gas-permeable than in other species.

Some species of fishes have become so specialised for air-breathing that their capacity for water breathing is impaired and they cannot meet all of their oxygen demand even in normally oxygenated water. They must breathe in air at least once in a while. Surprisingly for a fish, they will "drown" if they are forced to stay in water. Among such fishes are the South American pirarucu *Arapaima gigas* (this fish can get huge, at 4.5 m and 200 kg, and 95% of its large oxygen demand is provided by airbreathing), the freshwater butterflyfish *Pantodon buchholzi* (popular in the aquarium trade), the adult *Lepidosiren paradoxa* (a South American lungfish), the electric eel *Electrophorus electricus*, the three-spot (=blue) gourami *Trichogaster trichopterus*, and the climbing perch *Anabas testudineus*. Some air breathers, like the European eel (*Anguilla anguilla*) or the trahiras (family Erythrinidae), are capable of overland treks that take them deep into forests.

Others can catch insects on land, like the afore-mentioned mudskippers (subfamily Periophthalminae), the mangrove rivulus *Rivulus marmoratus*, the reedfish *Erpetoichthys (=Calamoichthys) calabaricus*, and the eel catfish *Channallabes apus*. Mudskippers and the eel catfish can even bend their head downward to facilitate the capture of prey lying on the ground, an anatomical adaptation – shared with the salamanderfish *Lepidogalaxias salamandroides* – that is exceptional among fishes. Other amphibious fishes, like the North African catfish *Clarias gariepinus* (= *lazera*) and the climbing perch, take advantage of their ability to live on land to include grain in their diet. Others, like the goby *Lentipes concolour*, can surmount waterfalls as high as 130 m (about 425 feet) by climbing on the rocks behind or next to the falls, using a special suction cup formed by the fusion of their pelvic fins to cling to the vertical surfaces.

They do this amazing feat not as adults, but as fry no more than 2.5 cm (1 inch) long! The mangrove rivulus lives in streams that are prone to drying up, so air breathing is a definite advantage. A recent study done in Belize and Florida has found that these fish can find out-of-water refuge in an odd place: the inside of rotting logs. These logs have long galleries inside them that were originally bored by terrestrial insects. The rivulus, which normally are very aggressive towards one another, agree to a truce and pack themselves tight inside those galleries. There they can remain, without food, breathing moist air, for up to 66 days.

When it comes to living out of water for a long time, the undisputed champions are the African lungfishes (four species in the genus Protopterus). As the swamp in which they live dries up, they dig a burrow, coil up into it, and secrete mucus around themselves to form a cocoon. Then they become dormant. Their metabolism is very low, yet they still need to breathe a little bit, and they do so with air. This estivation goes on for the whole of the African dry season, and we're talking 7-8 months here. Even more impressively, some estivating African lungfishes have been forced experimentally to remain in this state for 4 years, and survived. The eggs of some fish can also survive in air, as long as the air is well laden with humidity. The splash tetra *Copella arnoldi* – itself an air-breather – can jump straight out of the water and deposit eggs or sperm on the underside of leaves. Then from below the males splash water onto the eggs every 10-15 minutes to keep them moist and oxygenated. In an aquarium, splash tetras can lay eggs on glass above the waterline, or even on the underside of the tank lid. Other species, even though they are not air breathers, are known to lay eggs in places that get exposed to air. For example, in Australia and New Zealand, several species of *Galaxias* deposit their eggs in vegetation above the waterline. The eggs hatch only when floodwaters rise high enough to submerge them. In Canada, populations of "white" sticklebacks lay their eggs in filamentous algae or even on bare rocks in the intertidal zone, places that get exposed to the air at low tide; exceptionally for a stickleback, the eggs receive no parental care. Sites chosen for terrestrial egg laying are naturally moist – or kept moist, as in the case of the splash tetra – so that the eggs won't dry up and die. By developing in air, eggs may benefit from higher temperatures, more oxygen available, and protection against aquatic predators.

Oxygen and Predation

As with aquatic surface respiration, air-breathing entails a substantial risk of predation because of the fish's necessary proximity to the surface. To demonstrate this risk of predation, Don Kramer and his students hand-reared a green heron until it was 1 year old and allowed it to forage in pools stocked with various kinds of fishes (10 species of tropical fishes, as well as bluegill sunfish and central mudminnows). The researchers experimentally reduced dissolved oxygen to either 1.6 or 0.5 ppm. As expected, the fishes came to the surface, albeit not as much as they would have if the heron had not been there – they were aware of the bird's presence and the danger it represented. Yet they were still forced to come up once in a while to perform aquatic surface respiration or air-breathing, especially at 0.5 ppm. The heron caught more fishes at 0.5 ppm than at 1.6 ppm because of the prey's greater use of the surface at that concentration.

The danger inherent to aquatic surface respiration and air-breathing was thus confirmed. With this risk in mind, it is worth mentioning that in many species of air-breathing fishes, individuals can synchronise the time at which they break the surface to gulp air. Presumably, the fishes are wary of predators and delay air gulping until they see another fish get away with it. The link between predation and synchronous airbreathing has been demonstrated experimentally by John Gee of the University of Manitoba. He simulated a heron strike by plunging the wooden model of a heron head into the water of a tank housing central mudminnows, *Umbra limi*, and saw that the mudminnows started to synchronise their air breaths after such a disturbance. Instead of each individual breaking the surface at any time as they did before, the mudminnows experimentally scared by Gee's fake heron tended to break the surface more simultaneously. Other students and colleagues of Don Kramer have investigated the risk of predation by piscivorous fishes rather than birds. Robert Poulin has found that guppies did not run a greater risk of being caught by oscars under hypoxia, probably because the predatory oscars were just as affected by the lack of oxygen as their prey. In contrast, Nancy Wolf has observed that giant snakeheads, *Channa micropeltes*, were more successful at capturing dwarf gouramis, *Colisa lalia*, at 1 ppm than at 3 or 8 ppm, because the gouramis had to leave the cover of submerged plastic plants to go near the surface.

The difference between the two studies may be that predatory air breathers, such as snakeheads, are not as adversely affected by hypoxia as predatory water breathers, such as oscars. Findings of this kind allow us to predict that oxygen-poor habitats could provide a refuge to hypoxia-tolerant prey species relative to their not-so-tolerant predators.

This idea has been put forward to explain the disappearance of indigenous species from the open waters but not from swamps in waters infested by the predatory Nile perch in Africa, and to explain the boldness of minnows in the presence of visibly distressed yellow perch under hypoxic conditions. When pursued by predators, tolerant species could do more than just seek hypoxic refuges. They could simply jump out of the water and spend some time on land. This predator-evasion behaviour has been observed in rivulines (family Aplocheilidae), killifishes (family Cyprinodontidae), and the bald sculpin *Clinocottus recalvus*.

Winterkill

In northern lakes, whole fish populations are sometimes wiped out during the winter. Following a fierce winter, very few fishes are found alive in the lake. Such massive die-offs – called winterkill – take place when the surface of the lake gets covered by ice, with snow on top. The ice isolates the water from the air, while the snow screens the aquatic plants from much of the already short daylight, curtailing photosynthesis. Oxygen eventually runs out, and the fishes start to die. They cannot use aquatic surface respiration or unrestricted air-breathing because of the ice. They are reduced to gathering around air bubbles trapped underneath the ice, as witnessed in the field by John Magnuson and his team from the University of Wisconsin. But this, obviously, can only be a short-term measure. The fish's only hope for long-term survival – beside physiological adaptations such as the use of anaerobic metabolism, is to congregate near the mouths of inflowing tributaries that discharge well-oxygenated water (if the current is strong enough, ice does not form over the stream, and oxygen can diffuse into it). Such streams, unfortunately, cannot always be found, hence the winterkill.

Bringing Oxygen to the Eggs

Another aspect of behaviour very much affected by low oxygen is parental care. Many species lay their eggs on the surface of rocks and fan them. Fanning is an activity whereby the parent maintains a flow of water over the eggs by rhythmically moving its fins near them. This water movement is necessary because without it the breathing eggs would use up all of the oxygen present in the layer of water that immediately surrounds them, and thus would eventually suffocate. The flow created by the parent guarantees that well oxygenated water is in constant contact with the eggs. Given this, one would expect fishes to fan more when dissolved oxygen levels are low.

This prediction is borne out. I have taken a battery-operated oxygen metre to tide pools where male threespine sticklebacks tended their nests full of eggs, and have found that lower oxygen levels coincided with higher percentages of time spent fanning by the fish. Dutch ethologists working in the lab have also been able to coerce male sticklebacks into fanning more by connecting a tube to their nest and pumping deoxygenated water through it. Higher fanning levels caused by low oxygen represent a great expenditure of energy by parental males, to the point that they may lose weight and be unable to successfully breed a second time after the first stressful attempt. This was recently demonstrated in common gobies, *Pomatoschistus microps*, by Jackie Jones and John Reynolds from the University of East Anglia. A peculiar behaviour has been documented in one species of cichlid, the rainbow cichlid *Herotilapia multispinosa*. Working in the lab of Miles Keenleyside at the University of Western Ontario, Simon Courtenay exposed parental rainbows to oxygen concentrations of only 2 ppm.

When the eggs hatched, the parents did not gather the emerging wrigglers into pits dug in gravel as usual. Instead they sucked the wrigglers into their mouth and spat them into vegetation, onto which the wrigglers attached themselves thanks to glue-producing glands on their head. Courtenay showed that this behaviour, called "wriggler-hanging", was more prevalent when oxygen concentration was low. Presumably, wrigglers benefited from being close to vegetation because of the oxygen released by photosynthesis and also because the plants were often closer to the surface. Wriggler-hanging has also been reported in other cichlids, such as the angelfish, the red discus, the severum and the festivum. One of my students has told me about another peculiar behaviour, this time in the convict cichlid. He had a pair of convicts tending their eggs inside a ceramic skull (one of those aquarium decorations sold in pet shops).

The female, instead of fanning her eggs in the usual way, used to go to the airstone, take air into her mouth, bring it back to the nest and release it at the bottom so that the air bubble would float upwards along the egg batch. She did this repeatedly, to the extent that air was gathering at the top of the skull, threatening to raise the nest and its content all the way to the surface! Did she carry air bubbles to provide oxygen to her eggs in the face of localised hypoxic conditions? George Barlow has provided another interpretation, which may be better: he has observed a similar behaviour in orange chromides, *Etroplus maculatus*, and he saw this as an expression of fry-retrieving behaviour. The small bubbles are about the same size as fry. The parent may be fooled into thinking that the air bubbles are fry (even though its eggs have not hatched yet), and it tries to bring back those incredibly mobile fry back into the nest.

There is another similar case, but one in which an oxygen-supply explanation seems more likely. Some gouramis take in air, move below their bubble nest, and with strong opercular action they squirt bubbles up through the nest. 31 This behaviour probably brings oxygen to the eggs and must have evolved as an adaptation to the warm, stagnant and therefore oxygen-poor waters in which these species eke out a living. The ease of ventilating eggs may be influenced by nest shape. In the common goby, males build a nest for their future eggs by dumping sand over down-turned mussel shells. In the lab they may do it over a clay flower pot. The accumulated sand partially occludes the object's opening. Normally the males try to make the opening relatively small because this impedes the entry of egg predators such as crabs.

But small openings also mean less potential for circulating water. It has been observed that the owners of nests with smaller openings fan more than others. Also, if the oxygen content of the water is experimentally reduced, then the males build nests with larger openings. Obviously the fish adapt their nest building and nest tending behaviour to oxygen levels. In many fish species, females prefer to lay eggs in nests that already contain other eggs, perhaps a form of copying the choice that previous females have made for particular high-quality males. But considerations relating to oxygen must be taken into account. In sticklebacks for example, females prefer to spawn in nests that already contain two or three clutches, but they disregard nests that are packed with four or five clutches. This may be related to the fact that eggs in an overcrowded nest have trouble getting enough oxygen for optimal development.

In the same vein, female common gobies normally prefer nests that already have eggs in residence, but if the fish are kept in oxygen-poor water the females reverse their preference and now favour nests that are empty. This new choice is probably adaptive because competition for oxygen is fiercer in a nest with many eggs, and such a nest should therefore be avoided when oxygen is already in short supply. The South American lungfish *Lepidosiren paradoxa* lives in swamps where hypoxia is common. In this species the male guards eggs and developing larvae within a nest. At such a time, the male grows long filaments on his pelvic and pectoral fins. Work in the 1930s has shown that oxygen is released from these filaments into the water. It is thought that the filaments act as oxygenators for the nest and its content. The male breathes air at the surface, oxygen passes into his blood, and from there it is released into the nest through the thin surface of the filaments.

Physiological Adaptations to Low Oxygen Levels

Adaptations to low oxygen can be not only behavioural but also physiological. Fishes that live in frequently hypoxic habitats may have more haemoglobin in their red blood cells, and more of those cells in their blood, and therefore a higher blood capacity to take up and transport oxygen. Their body tissues may contain more myoglobin, a molecule that can bind up oxygen and therefore act as an oxygen store. But their main adaptation is anaerobic metabolism, a set of biochemical pathways that do not require oxygen to yield energy. This type of metabolism is not very efficient and can lead to the accumulation of relatively toxic by-products, such as lactic acid, and therefore when oxygen is present anaerobic metabolism is put aside in favour of its more efficient aerobic counterpart. But when oxygen is rare and metabolic demand is low, as in a cold water fish for example, anaerobic metabolism can contribute to survival for days, weeks, or even months. For example, through the use of anaerobic metabolism, goldfish can survive for up to 9 days at 4 oC in only 0.5 ppm of oxygen.

Similarly, from February to April there is virtually no oxygen at the bottom of northern lakes, and yet crucian carp, *Carassius carassius*, survive there because of their anaerobic metabolism and the cold winter temperatures that lower their energy requirements. Finally, drought is another ecological condition that selects for anoxia tolerance via anaerobic metabolism. For example, the killifish *Austrofundulus limnaeus* lives in ephemeral ponds in Venezuela and the eggs it produces can enter diapause and survive for up to 60 days in the complete absence of oxygen. Last word Behavioural and physiological responses are not sufficient to allow fishes to live indefinitely in severely hypoxic waters, but they can contribute to survival for significant periods of time. It is a comforting thought, one to which fish keepers can cling next time they are away from home on an extended leave, worrying about their air pumps suddenly failing.

How do Parasites Affect Fish Behaviour?

For most people in their right mind, parasites are disgusting little creatures. The mention of them evokes images of tiny crawlers entering the body of innocent victims by devious means, encrusting themselves within various organs (often with very unpleasant effects), living off their host's tissues, and then exiting the body, sometimes through scatological avenues, leaving behind hosts that are incapacitated, if not mere shells of their former selves (as is literally the case for some insects). The true facts are not always that gruesome (after all, parasites that have to spend a long time inside a host do not benefit if this host kicks the bucket, because then they die as well) but often enough parasitic reality is not pretty. Ask biology students about their parasitology course, and they will tell you that attending class is like visiting a little shop of horrors. I, for one, came out of my parasitology course with newly developed phobias (mercifully temporary) of such innocuous activities as eating meat and bathing in streams – they represent ways of transmission for some parasites. But students will also tell you that parasites are fascinating because of the complexity of their life cycles and the intricate ways in which they complete these cycles. So fascinating, in fact, that many students eventually become parasitologists and devote their entire lives to studying the little critters. Most of these people work with parasites of livestock or humans, but a few with an equal taste for ethology and ichthyology pursue the more esoteric questions that occupy this page: how fishes avoid getting parasitized, and how fishes alter their behaviour once they get parasitized.

How Fishes Avoid Getting Parasitized

This question has not been studied very much, and the few experiments I am aware of have all dealt with bloodsucking ectoparasites. These small invertebrates can be free-living, but when they perceive a passing shadow or some water disturbance, they jump. With luck, they land onto the fish that created the disturbance. They fixate themselves onto the skin or the gills and then they start their dirty work, sucking blood like miniature versions of lampreys. Because of this sit-and-wait strategy, ectoparasites resemble predators, and we would expect that any means of avoiding predators (shunning dangerous areas, shoaling more, moving less) should also work against these parasites.

This, in essence, is what the studies have revealed. Almost all of this work was carried out in the laboratory of Gerry FitzGerald, using populations of three-spined and blackspotted sticklebacks that were plagued by the crustacean ectoparasite *Argulus canadensis* (the Canadian fishlouse). Free-living parasites, along with infected and non-infected sticklebacks, could be captured in the field and brought back to the lab. A doctoral student, Robert Poulin, noted that when one half of his tanks was decorated with plastic plants while the other was left open, the fishlice preferred to stay near the bottom in the "vegetated" area. Sticklebacks also preferred to swim near the bottom in vegetated areas but only when parasites were absent. When free-living fishlice were present, the sticklebacks' choice of habitat changed: they spent more time near the surface and in open areas, presumably to avoid the more infested sectors of the tank. Poulin then investigated another strategy for avoiding these parasites: shoaling more. He released 30 sticklebacks in circular wading pools that might or might not contain free-living fishlice.

He reported that, as compared to the absence of parasites, fish in the presence of lice spent more time shoaling and formed larger groups. The benefit from bigger shoals derived from a higher probability that the free-living parasites would be detected and eaten by shoalmates – fishlice cannot infect a fish from the inside and can therefore be consumed with impunity. Another benefit was a dilution of the risk of being targeted for an attack by any given parasite. Poulin then switched to another system (and another lab) and made the following observation: Within experimental tanks, brook trout fry that ended up getting infected more quickly by ectoparasitic copepods were also those that moved more within their environment. This result suggested that reduced movement is another possible strategy for avoiding ectoparasites. Later, back in FitzGerald's lab, Lee Dugatkin wondered if sticklebacks would avoid the company of parasitized conspecifics, given that fishlice can jump from fish to fish. He divided a long aquarium into three sections with transparent partitions. He put 10 healthy fish on one side and 3 healthy fish together with 7 infected ones on the other.

He then repeatedly put single fish in the middle and recorded which group they approached most often. By and large the sticklebacks spent more time close to the healthy school, probably because the infected fish on the other side displayed odd behaviours such as swimming erratically and rising to the surface. (The avoidance, therefore, probably was not to external signs of parasitism per se, but to conspecifics that behaved abnormally. Nonetheless, the end result is that parasite transmission is reduced) Even in the case of ectoparasites that cannot be transmitted directly from fish to fish, it may pay to avoid the company of parasitized shoalmates. This is because the abnormal behaviour or appearance of infected fish may draw the unwanted attention of predators that could then endanger the whole shoal. One workable system here is the association between the trematode worm *Crassiphilia bulboglossa* and its host, the banded killifish *Fundulus diaphanus*. This trematode encysts itself in the muscles of its host, and this cyst forms a black spot that can be seen externally through the skin. Jens Krause showed that, in choice tests, banded killifish preferred to associate themselves with completely unparasitized shoals rather than completely parasitized ones (surprisingly however, Krause's killies did not prefer unparasitized shoals over partially parasitized ones). The same experiment also revealed that, in this case, it is the black spot caused by the worm's presence that betrays infection to the eyes of shoalmates. When members of a shoal were injected subcutaneously with a speck of black ink, they still displayed the normal behaviour of healthy fish but were nevertheless shunned by their shoalmates in choice tests.

How Fishes Alter their Behaviour Once Parasitized

This brings us to the second question, that of altered behaviour in parasitized fish. The underlying theme here is that often fishes are not the only host in a parasite's life. An endoparasite's life cycle often starts with the first larval stage entering the body of an aquatic invertebrate such as a snail. From the snail a new larval form is transmitted to a fish, either through ingestion of the snail or directly through water. Once inside the fish, another larval stage of the parasite encysts itself. Then the fish gets eaten by another fish or by a bird, the final host, inside of which the adult parasite develops. The eggs of the adult parasite are then passed on in the predator's feces, hopefully over water in the case of a bird, and from there infection of a snail can start the cycle anew. So, once inside a fish, it is in the parasite's interest to somehow alter the behaviour or appearance of this fish in a way that increases its chance of being captured and eaten by the next host in the cycle. Most studies have looked for evidence of this so-called "manipulation hypothesis". The most convincing results to date have come from a study by Kevin Lafferty and Kimo Morris, working with the Pacific (=California) killifish *Fundulus parvipinnis* and its two trematode parasites, *Euhaplorchis californiensis* and *Renicola buchanani*. The researchers put wild-caught killifish in a 150-liter tank, selected focus individuals, measured their behaviour, and caught them again in order to dissect them and count the number of parasites inside. They found correlations between the level of infection and the frequency of conspicuous behaviours such as flashing, contorting, shimmying, and coming to the surface.

The heavier the infestation, the more common the odd behaviours were. Lafferty and Morris then tested the idea that such behaviours could make the fish more susceptible to predation. First, they captured killies from two different populations, one known to be parasitized and the other never parasitized. Then they placed known numbers of these fish in outdoor pens accessible to various species of wild herons, gulls, and fish-eating ducks. After 20 days, only 4 out of 53 unparasitized fish had disappeared (that is only 8%), as compared to 51 out of 95 parasitized ones (54%). Disappearance was caused by predation and not by other factors such as disease or escape from the pens: the study included a control experiment with enclosures covered by netting that prevented bird attacks, and almost no fish from either population disappeared from these pens. Lafferty and Morris concluded that parasitized fish are more susceptible to avian predators, to the benefit of their parasites.

How the parasites alter the fish's behaviour is unclear, but the fact that one of the trematode, *Euhaplorchis*, is found in the brain suggests that the production of some crucial neurotransmitters may be manipulated by the parasite. Indeed, a recent study has found differences in the expression of serotonin and dopamine in infected fish as compared to healthy ones. Other studies have explored parasite-host systems that reveal different mechanisms of host manipulation. For example, the common dace *Leuciscus leuciscus*, when heavily parasitized by the eyefluke *Diplostomum spathaceum*, has been found to spend more time in the top 10 cm of an aquarium, possibly because the eyesight of the fish is impaired by the parasite, causing a preference for the better-lit surface waters. The ultimate hosts of this parasite, gulls, are known to be more successful at catching fish that are closer to the surface.

The same parasite can also affect the eyesight of rainbow trout, reducing the fish's ability to match its colour with that of the environment, thus making them more visible to predators, and reducing also their reaction to moving objects – some of which might be predators. Another commonly studied system is that of sticklebacks and the cestode *Schistocephalus solidus*. These parasites live in the peritoneal cavity of sticklebacks in such great size and numbers that they sometimes cause a visible distension of the abdomen. The parasites divert for their own purposes much of the energy assimilated by the fish. Afflicted fish are therefore hungrier, and we know that hungrier fish are more willing to take risks in order to keep on foraging. Not surprisingly therefore, several studies have shown that parasitized sticklebacks, after being scared by the strike of a model heron bill, flee over shorter distances, remain motionless for shorter periods, and come back to a food patch sooner than healthy individuals. They are also willing to risk looking for food at the bottom of tubes that are located fairly close to a big predatory cichlid (behind a clear partition), something that healthy fish will not do.12 In a similar vein, parasitized nine-spined sticklebacks that perform aquatic surface respiration in hypoxic waters come back to the surface sooner than nonparasitized fish after being frightened by the passage of a model kingfisher over the aquarium. Infected fish may have increased energetic demands, requiring more oxygen.

Hungrier fish are also known to form shoals that are smaller, less compact, and less cohesive, attributes that favour foraging but make the shoal more vulnerable to predation. In addition, hungry fish spend more time at the frontal periphery of the shoal where they can be the first to find food, even though they may also be the first to encounter predators. Do parasitized fish show the same tendencies? Many studies have revealed that they do. Sticklebacks with cestodes, European minnows with helminthes, fathead minnows with flukes, as well as killifish with trematodes, leave more room between themselves and their neighbours, and stay at the front or periphery of the shoal – or leave the shoal altogether – more often than their healthier counterparts. In general, the method used in these studies was to set up a shoal, select a focus fish that was obviously parasitized (as betrayed by abdominal distension or by black spots on the body), measure its social behaviour, and compare it to that of a healthy individual within the same shoal.

That method is not as elegant as the one used by Lafferty and Morris. As described above, these two researchers identified some fish by external marks unrelated to parasitism, measured their behaviour, and dissected them later to learn of their parasite load. This is an example of what we call a "blind" experiment because the observers did not know in advance what category (heavily parasitized, slightly parasitized, unparasitized) the focal fish would belong to. Therefore, they could not be biased while scoring the fish's behaviour. Observer bias – an unconscious tendency to measure ambiguous behaviour in a way that will make the results fit the expectation – is a constant danger in ethological studies. Researchers must compensate (but not overcompensate!) by trying to stay conscious of this danger or by using blind protocols.

Parasitism and Mate Choice

It may sometimes fit the ethologist's purpose not to know in advance the level of parasitism in fish, but the same cannot be said of the fish themselves. Females looking for a mate, for example, probably want to know if the prospective father of their young will have the stamina to care for the eggs and fry (in the case of parental species), and stamina is generally curtailed by parasitism.

Females would also prefer males that could pass on to their offspring whatever heritable trait they may possess for resistance to parasites. Females would therefore be looking for a body feature that could reveal the male's infection level. By choosing to mate with males that have such features, with a preference for those males that show marks of health, females could create a selection pressure for the existence of sexual ornaments that reliably divulge the male's level of parasitism.

This hypothesis was first published in 1982 by William Hamilton and Marlene Zuk, originally in the context of bright colours in birds, but tested a few times with fishes since then. The most striking of these tests was carried out by Manfred Milinski and Theo Bakker at the University of Bern. They started by placing two male three-spined sticklebacks next to a tank containing a female that was ready to spawn, and noting how much time the female spent near each male. They found that she usually spent more time close to the male with the brightest red colour on his belly. That this preference was based on colour and not on some other correlated characteristic (such as courting intensity, for example) was elegantly demonstrated by offering the same choice of two males under green light, which rendered the red colour uniformly black, and finding that the female then chose at random.

Milinski and Bakker next offered a choice between males that had never been parasitized versus males that had been recently infected by a ciliate, *Ichthyophthirius multifiliis*, a parasite that impairs body condition and reduces the intensity of body colours. By the time of the tests, the external white spots caused by the ciliates had dropped off the fish, but the red colouration of the formerly parasitized males was still duller than that of their unparasitized conspecifics. As expected, the female sticklebacks spurned the dull males under normal lighting, but not under green lighting. Avoidance of this kind can work both ways, and in some species the tables are turned on the females. In pipefishes for example, if females show external signs of parasitism such as black spots – either natural ones or artificial ones that have been created by injecting ink underneath the skin – males spend less time next to them.

Other Effects of Parasites

Parasite infection can be correlated with behaviours that do not seem to give rise to increased predation risk. Some of these effects suggest a debilitating effect of parasites, but others are harder to interpret. For example: Salmon harbouring helminth parasites seem to orient less well during their seaward migration. Infected minnows are less well synchronised with their shoalmates. Upland bully *Gobiomorphus breviceps* and guppy females infected with various parasites have problems discriminating between males of different quality and size. Parental male three-spot damselfish, *Stegastes planifrons*, experience more egg loss when they are parasitized, though it is unclear if this is due to a lesser ability to protect the nest from predators or from parasite-induced hunger which leads to egg cannibalism by the father. Sticklebacks blighted with *Schistocephalus* (them again!) generally prefer to eat smaller prey. Perhaps the infected sticklebacks cannot compete with other fish for the larger food items, or maybe the overabundant internal parasites constrict the stickleback's stomach so much so that only small prey can be consumed. On the other hand, though sticklebacks parasitized with *Schistocephalus* consume smaller prey items, when they have an ample food supply and no competition they seem to eat a lot of those small preys, as they end up growing *faster* than non-parasitized fish, even when the weight of the parasites is subtracted. It is unclear whether this is a response of the host fish to try and better marshal defences against the parasite, or a manipulation of the host's physiology by the parasite to make it more visible to predators. Faster growth leads to bigger body size, and bigger fish are more visible to predators.

Chapter 3

The Sex Lives of Fishes

In fishes as in all living beings, reproduction is the biggest game in town. The behavioural, anatomical, and physiological features that we see in fishes today are there because they were successfully passed on from generation to generation throughout the ages. They were passed on in the form of genes (bits of DNA) during the reproductive act. Those individuals that did not care about reproduction died without transmitting their genes to the next generation, and their contempt for the reproductive enterprise died with them. In contrast, those individuals that had an innate taste for mating propagated that trait through the population. Evolutionarily speaking, stakes are very high in the reproductive game. There is pressure on all individuals to leave as many viable descendants as possible in the next generation. One way of doing this is to choose a sexual partner of high quality, one who will impart genetic value to the offspring, as well as taking good care of them. But for those individuals who score low in the mate choice game, there are other ways to pass on their genes, less flashy ways, tactics that are less straightforward, some might say less fair-play, but reasonably efficient nonetheless. These are the dirty little secrets of a fish's sex life.

Kleptogamy (Sneaking)

In almost all fishes, fertilisation is external. The male's sperm and the female's eggs are released from the body and they must meet in open water for fertilisation to occur. While they float in water, a female's eggs are vulnerable to interception by the sperm of a male who may not be her chosen mate. Welcome to the world of kleptogamy, the stealing (*klepto*) of fertilisations (*gamy*) by an interloping male. Sometimes the cuckolding male is a territorial neighbour. After realising that a female is about to spawn next door, this neighbour tones down his bright nuptial colours (which would identify him as a territorial rival), becomes more female-like in appearance, and sneaks into the courting male's territory. When the female sheds her eggs, the sneaking male rushes to her and releases his sperm. Meanwhile the rightful mate, who himself might be in the act of spawning, is caught unawares. He usually gathers his wits and chases the intruder away, but often the damage is already done. Some of the sneaker's sperm has mingled with the female's eggs, resulting in at least some illicit fertilisations. Such a scenario is often played out in sticklebacks, sunfishes, and various cichlids.

In a great number of species (salmon, sticklebacks, sunfishes, minnows, darters, cichlids, wrasses, parrotfishes, gobies, among others) cuckolding part-time while holding a territory full-time does not suit some males. They prefer instead to forego territoriality altogether and to specialise in cheating. Their tactic is the same: they hide near the periphery of a spawning territory, watch what is going on there, sneak in at the critical moment, and try to release sperm as close to a spawning female as possible. This behaviour has been documented and studied for more than 30 years now. The parasitic males have variously been called sneakers, streakers, cuckolders, hiders, furtive males, accessory males, type II males, or interference-spawning males. Because they are small and superficially resemble females (all the better to fool territorial owners), they have also been dubbed pseudofemales, female mimics, and even transvestite males. They may swim around with authentic females, or hide near the nest or territory of the so-called "bourgeois" males, waiting for their chance to steal fertilisations.

Sneakers are parasites. They let bourgeois males do the hard work of defending a territory, building a nest, and developing body ornaments to attract females. Then, seemingly at little cost to themselves, they reap the benefit of at least some fertilisations. If the territory owner provides parental care, the sneaker's progeny will benefit from that too, for bourgeois males cannot tell which eggs were fertilised by them and which were not, and they take care of all eggs indiscriminately. But there are costs to being a parasitic spawner. Sneakers are pariahs of the high reproductive society. They are despised both by gravid females and by territory owners. In other words, females do not choose them as mates because they are usually too small, and males often recognise them for what they are, despite their disguise, and mercilessly attack them at every opportunity. In salmon, sneakers may even get killed by big males. This makes for a rather stressful existence for the little guys. Indeed, one study on the Mediterranean wrasse *Symphodus ocellatus* has established that sneakers lose just as much weight throughout a reproductive season as territorial males do.

Sperm Competition and Sperm Economy

Sneakers make a special investment in their unconventional way of life. They develop huge gonads. The testes of sneakers can be up to seven times the size of the norm for the species. This is because sneakers practice sperm competition. Their behavioural involvement on the spawning scene is obviously very short; they are in and out. They can only let their sperm do the talking. But the more sperm they release, the greater their chance of grabbing a good share of fertilisations. So, to produce more sperm, sneaker males invest into the development of big testes. (In group spawners – fishes, like herring and cod, that gather in groups and shed eggs and sperm simultaneously and indiscriminately – males also invest in large gonads, which makes sense because in a group spawning sperm from one male are engaged in a race with sperm from many other males.) 8 Bourgeois males, in contrast, save their strength for territory defence and invest less energy into sperm production. In fact, some of them can be downright stingy with their sperm. The first thing to know here is that male fishes seem to be able to control the number of sperm (the size of the ejaculate) they release during a spawning event. The second thing is that they appear to exert this control in a way that makes sense depending on the circumstances. In some coral reef species for example, a male may need to spawn with many females, sometimes in the same day, and it has been found that males do not apportion sperm equally between all females. With larger females, which usually produce more eggs, males release more sperm. With smaller females, males are less generous. The males do not deign to release more sperm than is necessary for a reasonable fertilisation rate, keeping the rest in reserve for future spawnings. Biologists call this "sperm economy". Another example of sperm economy can be found in species where males defend a territory that encompasses the domain of many females. These females become the male's harem. It has been found that the larger the harem of a male is, the fewer sperm he releases when spawning with any single female. He wants to make sure that he has enough sperm to go around the full harem, and this means smaller individual contributions when the females are more numerous. (If readers are wondering how researchers manage to obtain information on ejaculate size in field studies, they should prepare to be surprised at how simple it is. Snorkellers hover over likely spawning sites and make sure they do not disturb spawning pairs. But as soon as the deed is done, the snorkellers descend, scatter the fish, and trap all of the milt and eggs – still visible as a white plume in water and still fairly localised – within a large plastic bag. This water is brought back to a boat, mixed, and sampled. The eggs and sperm are stained and counted under a microscope. The tally is multiplied by the dilution factor of the sampling to obtain a final count.) Males can also adjust the volume of their ejaculate depending on the perceived risk of competition from other males. For example, when mating with a female a male mosquitofish *Gambusia holbrooki* will invest a greater proportion of the sperm reserve in his testes if he has previously been kept in the presence of another male (a potential competitor) as compared to being housed with females only. Male threespined sticklebacks will also invest more sperm during mating if a TV next to their aquarium displays the image of another male courting as opposed to another male fanning his eggs (parental stickleback males as a rule do not try to steal fertilizations and therefore represent a lower risk of sperm competition). Male bitterlings, *Rhodeus sericeus*, release more sperm into the mussels where females have laid their eggs if there are other males around, at least up to a certain extent. But the response to competition is not always sperm warfare: in some species, overt aggression and more attentive mate guarding is a behavioural alternative.

Success of Sneakers

Back to our sneakers: they may dominate in the arena of sperm competition, but they need this nominal advantage because bourgeois males start the game with a big ace up their sleeves. Bourgeois males are *with* the spawning female, next to her right from the start. They can better judge the exact moment when she will release her eggs. They can manage to be as close to the female as possible at that moment. They can synchronise the release of their sperm with the actions of the female. In terms of temporal synchrony and spatial proximity, bourgeois males outperform sneakers. So what is the end result? Who wins the majority of fertilisations? Researchers can answer this question by capturing all of the actors (males, female, fertilised eggs) on the scene of a spawning event and conduct paternity analyses. Results vary from event to event and from species to species, but as a whole it seems that sneakers almost never fertilise more than half of all the eggs. In fact, they usually do no better than a 33% fertilising rate. And of course, their success is nil when they happen to be away from their station – fed up with being chased all the time. However, these numbers are still high enough on average to make sneaking a viable tactic. In some populations of sunfishes, parasitic males may even outnumber territory owners. In a few species such as salmon and sunfishes, sneaking appears to be an innate habit, ingrained at birth and fixed for life for the individuals concerned, in the same way that territoriality is the only option for the other males. In the majority of species however, the reproductive role of males is conditional. Depending on the circumstances, the same male may either try his luck as a sneaker or invest into the maintenance of a spawning territory. The incentive to become a sneaker seems to be greater when the male is small relative to the rest of the population, when he is less experienced, and when there are many territorial competitors. (Rather than describing sneakers as behavioural parasites, it might be more charitable to view them as poor sods who are trying to make the best of a bad job; they gamely try to play the poor hand they have been dealt). In a few species, males have been known to switch back and forth between the two tactics, depending on the conditions that prevail at the moment.

Satellites

In a few species, there is an intermediate tactic between sneaking and holding a spawning territory. It consists of defending a small territory, or at least a particular station, at the edge of a big male's domain. Practitioners of this middle-ground philosophy are called satellites. They are more or less tolerated by the big males. They may breed with small females. They may also steal fertilisations, like sneakers, from the large males. On the other hand satellites can also attack intruders, including other sneakers. Therefore they occupy a buffer zone and form a first line of defence around the big males' territories. One of the best examples of satellites is in pupfishes, particularly the Pecos pupfish *Cyprinodon pecosensis*, where males can either be territorial bourgeois, brightlycoloured satellites, or drab sneakers. There are other instances in the cichlid, wrasse, and sunfish families.

Forced Copulations

Sneakers circumvent a female's mate choice by surreptitiously partaking in the chosen mate's spawning. Another tactic, rare but nevertheless readily observed in some species, is to somehow mate with a female privately but against her will. In livebearers, fertilisation is internal. The male deposits his sperm inside the female by inserting his modified anal fin (the gonopodium) into the female's genital pore. Normally, insemination is preceded by a display on the part of the male and acceptance by the female. But some males also try sometimes to approach a female from behind and thrust their gonopodium at her genital pore without prior display. Females normally try to evade such a sneaky manoeuvre, but they don't always succeed. On average however, the success enjoyed by thrusting males is low because of the female's lack of co-operation and the fact that the few sperm injected are often displaced later on by full copulations from displaying males. Yet, for some out-of favour males this low success is better than nothing.

In livebearers such as sailfin mollies, the swordtail *Xiphophorus nigrensis*, the Gila topminnow, and the mosquitofish, females shun small males when they choose a sexual partner. Small males compensate for this by resorting to thrusting. Careful observations of the social life of these species have revealed that big males display in front of females more often than they thrust, while small males attempt to thrust more often than display.

In another livebearer, the guppy, there seems to be no relationship between body size and thrusting frequency. Instead, thrusting becomes more prevalent when there is a risk that predators could spot displaying males or when there are more competing males in the environment. Moreover, under certain conditions (low light), thrusting predominates in those males that possess longer gonopodiums (be it absolute length or length relative to body size). This suggests that more sperm may be delivered when a longer gonopodium is used in thrusting, and that the propensity for sneaky fertilisations may be co-evolving with gonopodium length. This hypothesis awaits further testing.

Here is another example of forced copulation, this time in a species with the more normal pattern of external fertilisation. In the bluehead wrasse *Thalassoma bifasciatum*, females prefer to mate with big males on traditional spawning sites, but to get to these sites the females must run a gauntlet of smaller non-territorial males. These males intercept the females, relentlessly pursue them, and repeatedly touch them. Sometimes this induces the females to release their eggs – against their true intent, we must presume. All of the small males present on the scene then shed their sperm in what can only be considered a group spawning. As in species where group spawning is the norm, small bluehead wrasse males have bigger testes and release more sperm while spawning than their larger territorial counterparts.

Threesomes

In a few fishes, gravid females are often attended to by two or three males simultaneously. These various males do not appear to differ in status, and they more or less tolerate each other. When the female spawns, the males release their sperm together. This pattern has been observed in some minnows, salmon, and pike. In many species of suckers (family Catostomidae), spawning often occurs in trios. Two males adjoin a female on either side and seem to press against her. All three fish release their eggs and sperm at the same time. It has been suggested that the pressure exerted in concert by the two males squeezes more eggs out of the female than would normally be released, but this hypothesis has not been tested yet.

Sex Changes

A diverse array of fishes, mainly marine species, can change sex. Observations of individually marked fish in tanks or in nature have clearly showed that some individuals start their adult life as males, and switch later to females. Or they begin as females, and then become males. This is indicated not only by a change of behaviour, but also by a transformation of the gonads as revealed by biopsies and dissection. Testes are progressively invaded by ovarian tissue. Or ovaries can develop a vas deferens and turn into testes. Among large animals, such capacity for sex change is almost a unique feature of fishes. Mammals, birds, reptile, and amphibians (with the exception of the reed frog *Hyperolius viridiflavus* 27) are incapable of changing sex in nature. Sex changes usually happen in species for which the transformation is not too taxing anatomically. Usually these fishes have no elaborate external genitalia, only a duct that leads to gonads. It is hard enough metamorphosing one kind of gonad into another – changing sex can take days, sometimes even weeks – without having also to develop gonopodiums and other accoutrements of the sort. Another characteristic of sex changers is that for them the influence of age, and therefore of body size, on reproductive success is not the same for both sexes.

Small individuals do much better if they are of one sex while large individuals make more babies if they are of the other sex. This latter point requires an illustration. Anemonefishes often live as male-female pairs within anemones. Because anemones are often widely scattered, it would be difficult for a male to find and spawn with many females during a single reproductive season. Likewise, a female would have a hard time courting many males. So male and female are usually stuck with each other within their little world. Now, because sperm is relatively cheap to manufacture, a small male can fertilise all of the eggs that even the biggest female could possibly produce. In spite of his diminutive size, a small male can also easily defend those eggs because of the protection afforded by the anemone. Therefore, no reproductive advantage can accrue to being bigger for a male. On the other hand, the bigger a female is, the more eggs she can churn out. Therefore it would be advantageous for both members of an anemonefish pair to have the smallest individual be a male and the biggest one be a female. This has resulted in an evolutionary pressure on males to turn into females as they grow bigger. And this is what happens in most anemonefish.

Species: males can change sex and become females. Elegant work by Hans Fricke has shown that it is not size per se which determines when a male anemonefish changes sex. It is the *relative* size of the two partners and the presence or absence of interactions between these partners. When a male twobanded anemonefish, *Amphiprion bicinctus*, was experimentally forced to live alone in an anemone, he usually did not change sex, or did so very slowly. However, as soon as this lone male was joined by a smaller male, he rapidly turned into a female, reaching the capacity to lay eggs within as little as 26 days. By contrast, if the lone male was joined by a larger male, it was this larger male who changed sex. Absence of a bigger female *and* presence of smaller male combine to trigger sex change in this species. Something similar can be seen in the nosestripe anemonefish *Amphiprion akallopisos*, where several small males may live together with one female in a particularly vast anemone. Among the males, only the largest one gets to breed with the female. All the others are "psychologically castrated" by the aggressive actions of the breeding male. If the female is experimentally removed, this large breeding male, and only he, immediately starts to turn into a female.

Once this is accomplished, the next largest male in line is promoted to breeding status. Sex change is anemonefishes is only from male to female, not the other way around. If the male of a pair disappears, the remaining female must patiently await the chance arrival of another male, or she must set off on a quest for a new partner (this latter alternative is dangerous, because anemonefishes are poor swimmers and away from the protection of their anemone they are easily captured by predators). Lone females never metamorphose into males.

Even when two females are experimentally forced to cohabit within an anemone, the smaller of the two does not turn into a male, even though she suffers constant harassment at the hands of her dominant roommate. Socially-controlled sex change also exists in the Clark's anemonefish *Amphiprion clarkii*. However, sex change in this species can sometimes be short-circuited by timely migrations. This occurs in dense populations where anemones are clustered together so that movement between anemones takes little time and entails little risk. If the female of a pair disappears, the male will often eschew sex change and go look for another available female nearby. This can work to his advantage because sex changing takes several weeks, whereas finding a new mate in a dense population can be a matter of days only.

Despite its common appearance in anemonefishes, male-to-female sex change is rare for fishes in general. More prevalent is the pattern of female-to-male change. This one is in fact fairly common in wrasses, parrotfishes, porgies, damselfishes, seabasses, and gobies. In those species, large males enjoy great reproductive success because they exert exclusive control over extensive harems of females, or they preside over popular spawning territories. This leaves a sizeable portion of the male population with little mating opportunity. Rather than content themselves with the marginal success afforded by sneaking, such small fish may prefer to start out their lives as females, because theoretically all females have access to reproductive males and therefore a guaranteed shot at producing many descendants.

Only when they reach impressive physical stature and become more competitive do the most dominant females turn into males, assuming that an opportunity presents itself (i.e., after the loss of a resident male or the sudden influx of many additional females). As with the male-to-female sex change of anemonefishes, female-to-male change seems to be triggered mostly by social factors. For example, in the cleaner wrasse *Labroides dimidiatus*, which lives in groups that normally comprise a single male and a large harem of females, removal of the male results in the most dominant female changing sex, unless a larger male from outside barges in and takes control of the group. Amazingly, if a larger dominant male appears *after* the largest female has already turned into a male, that female-turned-male can change back into being a female. (Such cases of sex change reversals, also called bi-directional sex changes, are exceedingly rare; to my knowledge, the only other example is the coral-dwelling goby *Gobiodon histrio* from the Great barrier Reef: when two females-turned-males are placed together on a reef, one of the two males can turn back into a female, allowing mating between the two.) In Hawaii, Robert Ross, George Losey, and Milton Diamond conducted the following experiments with the saddle wrasse *Thalassoma duperrey*. They placed a number of lone females in individual pens; all of those fish stubbornly remained female.

However, if one of those lone females was joined by a smaller female (or, interestingly, by a smaller male), then she changed sex and turned into a male. If she was joined by a larger female, or by both a larger and a smaller female, she did not change sex; the joining large female did. These results were obtained even when the newcomers were separated from the original occupant by a transparent partition that allowed visual contact but no tactile exchange. The effects, however, could not be obtained when the partition allowed water flow (chemical exchange) but not visual contact. The researchers concluded that through visual cues only, presence of a smaller conspecific *and* absence of a large one can trigger sex changes from female to male in this species.

Sex Determination

It is worth mentioning here that in many fishes, sex is not determined at conception or at birth – there are no sex chromosomes. In such species (for example, silversides, Apistogramma cichlids, some livebearers), all individuals come into this world with undifferentiated gonads, or with gonads that look more like ovaries. Only later in life do the gonads become true ovaries or take a left turn and become testes. Environmental factors may affect the direction in which this development proceeds. In the Atlantic silverside *Menidia menidia* for example, low temperatures produce more females while high temperatures favour the formation of males (as is also the case, by the way, in alligators and many lizards). Essentially this means that more females are produced early in the spring, when temperatures tend to be colder, and more males are born later in the summer.

Thus females benefit from a longer growing season which allows them to make more eggs. Males manage to manufacture the (cheaper) sperm they need even though their growing season is shorter. In zebrafish *Danio rerio*, very low oxygen concentration (0.8 mg/L) affects the synthesis of sex hormones and results in slightly more males being produced (about 74 %, as compared to 62 % in normal populations). The extra males are genetically female but have testes instead of ovaries. More work is needed to find out whether the hypoxic conditions that prevail in the hypoxic, so-called "dead zones" of the world's oceans masculinize the fish populations that manage to live there. In some cichlids and swordtails, extreme pH can nudge gonad development along specific pathways, a very low pH yielding more males.

In some cichlids of the genus *Apistogramma*, both temperature and pH can determine sex ratio, with a critical period of exposure 30-40 days after birth. 40 There is even some evidence, at least in Midas cichlids, *Amphilophus citrinellus*, that relative size and social interactions can determine sex. In the company of smaller conspecifics, Midas cichlid 6-12 months old tend to become males, while other individuals of the same age placed with *larger* tankmates become females. In eels of the genus *Anguilla*, which includes the American eel *Anguilla rostrata*, more males issue from populations that are crowded. Low population density results in the predominance of females. In juveniles of the rippled coralgoby *Gobiodon erythrospilus (= rivulatus)*, which can be found on Australia's Great Barrier Reef, gonads do not mature until the fish has met a potential sexual partner.

Such a meeting is not a common event, because all the real estate on the corals tends to be already occupied by breeding pairs, and because extensive cruising is discouraged by the presence of many predators (coral gobies are small and make good snacks). The amazing thing is that the encounter with an available adult (most likely one whose mate has just died) will trigger sexual maturation by the arriving juvenile into whichever gender complements the new partner: male if the juvenile has found a female, female if the new partner is male. When it comes to pairing up, anygoby will do, thanks to flexible socially-influenced sex determination.

Sexual Maturation

Sexual maturation (gonads becoming capable of producing sperm or eggs; puberty, so to speak) usually occurs once a certain body size is attained. But in some case, social conditions can force a delay. The coralgoby above provided an example: sexual maturation had to wait until a partner was found. There are other examples of social control, and they almost all concern delays caused by the presence of adults. In several species of poeciliids, cichlids, and labrids, as well as in the characin *Corynopoma riisei*, juveniles raised with adults are inhibited from maturing as compared to juveniles raised alone. Usually this phenomenon takes place in males, though in the wrasse *Pseudolabrus celidotus* it is the female whose maturation is inhibited. Maturation is not delayed forever however: eventually the juveniles reach such a large size that they do become sexually mature. But the juveniles raised alone mature earlier, at a smaller size.

Social inhibition of maturation can be achieved either through chemical cues emanating from adults, or by the mere sight of sexually active males, or direct aggression by the more dominant adults. Social inhibition may be seen as manipulation by the adults to decrease competition for matings, but perhaps the response is adaptive for the juveniles as well. If there are lots of adults to compete with, it may pay to delay maturation and invest energy into more growth so that, when maturation does take place, the emergent adult will be sufficiently large to secure a fair share of matings.

Hermaphroditism Hermaphrodites play on two fronts: they can produce both eggs and sperm at the same time. (Biologists call this pattern "simultaneous hermaphroditism". The expression "sequential hermaphroditism" also exists; it refers to sex changers.) No vertebrates beside a few fishes are known to be simultaneous hermaphrodites. The best-studied fish group is a subfamily of sea basses (the hamlets, Serraninae). In these fishes, testes and ovaries are both present within the body but they are separate and they connect to the outside via different ducts. This arrangement precludes internal self-fertilisation. Individuals in fact mate together as pairs, one member playing the role of the female and releasing a few eggs while the other plays the male part and sheds a few sperm.

The roles are reversed a few moments later, and this quick alternation between male and female roles continues throughout the spawning episode. This behaviour is called "egg trading". It may have evolved to prevent individuals from cheating and specialising into the production of sperm only, which is cheaper to manufacture. Such a cheat could not have a high reproductive success because spawning between he and the hermaphrodites would quickly break down once the hermaphrodites realised he could not produce eggs and participate in the egg trading ritual.

The barred serrano *Serranus fasciatus* presents a mating system that incorporates several of the elements covered so far. In this species, all fish start life as hermaphrodites, but in most groups the largest individual loses female function and becomes exclusively male (a sex change of sorts). This male maintains and defends a spawning territory around which the other fish in the group (all of them hermaphrodites) gravitate. As females these hermaphrodites mate only with the big male. As males the hermaphrodites try to sneak and steal fertilisations from the big male. Few hermaphrodites can self-fertilize. The only well known case is the mangrove rivulus *Rivulus marmoratus*, a tough coastal dweller that can withstand heavy pollution, low oxygen levels, brackish water, and even the lack of water. It colonizes small streams that are prone to drying up.

Pure females don't exist in this species. About 95% of a population is born hermaphroditic, the rest being male only. At 3-4 years of age, about 60% of the selffertilizing hermaphrodites lose their female function and become strictly male. This transformation takes place only if the water temperature is below 20 °C. Above 25 °C, all fish remain hermaphrodites. Self-fertilization by the hermaphrodites may ensure reproduction in environments that are harsh and where other members of the species are rare. On the downside however, as opposed to ordinary sex, self-fertilization does not generate much genetic diversity. The presence of males in some populations of otherwise self-fertilizing hermaphrodites suggests a selective pressure to maintain a minimum of genetic diversity.

Parthenogenesis (Males, who Needs Them?)

Some species of fish are parthenogenetic. They are all females. They reproduce asexually. However, some parthenogenetic females still need males, in a roundabout way, from closely-related species. Eggs from these females develop only after being activated through contact with a male's sperm. However, although the male's sperm initiates the development of the eggs, it contributes no genetic material. The activated eggs invariably develop into new females. This particular brand of parthenogenesis is called gynogenesis. The best known example is the aptly named Amazon molly *Poecilia formosa*, with the closely related sailfin molly *P. latipinna* and shortfin molly *P. mexicana* acting as trigger males.

It may be asked why male sailfin and shortfin mollies bother to shed sperm over eggs that will not take in their genes. The answer may lie in a study that has showed how female sailfins prefer males they have seen consorting with other females, even when those females are Amazons (in many species, including mollies, females prefer to mate with males they have seen in the company of other females, a phenomenon called "mate choice copying"). The advantage to males of mating with Amazon mollies is that it makes them look good in the eyes of females of their own species. However, there are limits to what these males are willing to do for Amazons: when housed with Amazon females, male sailfins produce less sperm than in the presence of females of their own species. They are less primed (to use the technical term), a physiological analog to sperm economy. Some female sharks have been known to give birth in captivity after being separated from males for extended periods of time. However, because female sharks can store sperm for a long time after being inseminated by males, there was always a nagging suspicion that the new mothers had mated with a male before capture and simply delayed giving birth. These doubts have now been put to rest, at least in one case. In December 2001, at the Henry Doorly Zoo in Nebraska, one of three female bonnethead sharks (*Sphyrna tiburo*) captured while they were juveniles and kept together for three years without any contact with a male gave birth to one young. Genetic analysis has since proven that no paternal DNA was present in the young, thus confirming parthenogenesis in this shark species. Another shark species, the blacktip *Carcharhinus limbatus*, has recently been confirmed as capable of parthenogenesis also.

Homosexual Behaviour

In guppies, a species that practice internal fertilization, males spend a lot of time engaged in sexual behaviour directed at female guppies. As already mentioned, they either perform the sigmoid display (curving their body in the shape of an S), which is an invitation to copulate, or they attempt sneak copulations (trying to suddenly thrust their gonopodium into the genital pore of an unwilling partner). These two behaviours are easy to recognise. Sometimes, it is possible to witness a male directing a sigmoid display or a sneak attempt at other males. A study by Kristin Field and Thomas Waite at Ohio State University in Columbus suggests that the absence of females in the social environment of a male early in his life can cause homosexual behaviour. Field and Waite raised male guppies for 15 weeks in all-male groups or in mixed-sex groups, starting right after sexual maturation in the fish. At the end of those 15 weeks, each male was placed in the presence of three other males and three females. Males raised in the mixed-sex group directed almost all of their displays and sneaky copulation attempts at the females. In contrast, males raised in the all-male groups directed three times as many sigmoid displays to the other males than to females, and almost all of their sneaky copulation attempts at the other males, even though females were present. The scientists went on to warn that conservation breeding programs (not necessarily in guppies but in other fish species) should be careful about housing fish in single-sex groups. Many cichlid fishes form mating pairs. A male and a female develop a bond and start preparing a nest site together and defending a territory around it. Eventually the female lays her eggs, the male fertilizes them, and both protect the eggs. In cichlids such as the orange chromide and the jewel fish, if only females are housed together, two individual females may pair up. One or both of them may lay eggs, which of course remain unfertilized. Such homosexual behaviour does not happen if mature males are also present in the environment. (In contrast, male homosexual pairs never form, even in all-male groups. Male cichlids are just too aggressive towards each other. Even in females, aggression puts up a formidable barrier to homosexual pairing.) Even within these species however, homosexuality does not appear to be notable under natural conditions.

Fishes are said to be "lower" vertebrates. Technically this expression only means that fishes developed in the course of evolution earlier than amphibians, reptiles, birds and mammals. Yet the adjective "lower" cannot help but to convey an impression of inferiority. This is unfortunate, because the notion of what is inferior and what is superior depends on the point of view. For example, as compared to birds and mammals, fishes have retained an amazing plasticity in the development of their sexual anatomy and sexual behaviour. This has opened to them several alternatives to the "traditional" sex roles of big competitive males and smaller choosy females. These alternatives – sneaky males who look like females, males who can control their output of sperm, sex changers, hermaphrodites – are not inferior. The proof of it is that after so much time since fishes first appeared on earth, these seemingly odd reproductive tactics are still around, enabling their adherents to contribute specific genes to the next generation.

Are Fishes Good Parents?

Parental care is not the rule in fishes. Most species are quite content to abandon their offspring to the vagaries of a world populated by organisms that are fond of eating nutritious little snacks such as eggs and fry. These non-parental fishes may show discrimination as to where the eggs are deposited (covered by gravel in salmon, in thick mats of weeds for some minnows, in open water downstream from the reef in some wrasses) but I would classify this as careful spawning, not parental care. Though they may form a minority, parental species are nevertheless very diverse. They belong to many different families, marine as well as freshwater. About 80% of these families are represented by species that care only for eggs. These fishes prepare or build nests, sometimes nothing more than a cleaned rock, sometimes an amalgamation of vegetal matter, empty shells, pebbles, sand, or even air bubbles. At this site the eggs are vigorously defended against potential predators. (Some yellowspotted triggerfish, *Pseudobalistes fuscus,* have sent divers to the hospital with serious bites to the legs 2). Parents also tirelessly fan the eggs – moving water over the brood with their fins – to provide them with a good supply of oxygen (eggs consume oxygen like most other living things, and if water was stagnant the layer next to the eggs would soon become depleted and no more oxygen would be available). In addition, the parents clean their eggs by brushing them with their fins. With their mouth they remove dead or diseased eggs. Some also transfer onto their eggs mucus that has anti-microbial properties. If the nest is in the intertidal zone, parents cover the egg batch with algae just before low tide, preventing desiccation. One species of characin even jumps onto a low-lying leaf above the water, lays and fertilises eggs that stick to the leaf, and then the male periodically splashes water from below onto the eggs to keep them moist. The young drop into the water when they hatch.

Many species carry eggs with them so that if a predator makes an entry onto the scene, the parents can flee and bring the brood with them. The eggs are carried outside or inside the body. Examples of outside-carriers include medakas (eggs stuck to the female's vent for a few hours), bagrid and banjo catfishes (eggs embedded on the surface of the belly) and suckermouth armoured catfishes (eggs attached to the male's lower lip). Inside-carriers include the numerous species of mouth-brooding cichlids, sea catfishes, lumpfishes, cardinalfishes, and gouramis (eggs carried inside the mouth, and regularly churned in there for proper cleaning and oxygenation), many seahorses and pipefishes (eggs developing inside a special ventral pouch on the male), and finally, all livebearers as well as many sharks and rays, where young develop inside the reproductive system of the female, sometimes nourished by special connections to a placenta-like structure. (In pipefishes too there are placenta-like connections between parent –always the father– and embryos, and nutrients pass up from father to embryos, but it can be a two-way street: if the father is malnourished, he can actually use his "placenta" to suck his embryos dry, a case of filial cannibalism.) In some sharks like the porbeagle and the salmon shark, embryos inside the female feed on additional eggs that are produced by the ovaries for their sole benefit. It is estimated that only 20% of parental fishes care not only for their eggs but also for the fry that hatch out of them. These species include most cichlids, which are well known because of their popularity in the pet trade business.

Cichlids that lay eggs on a substrate (as opposed to those that carry eggs inside their mouth) are biparental. Both the male and the female care for the eggs and young. Outside of the cichlid family, such gender equality is rarely seen in parental fishes. In cichlids it is generally assumed that both parents are needed to defend the mobile fry against predators. Cichlids breed in areas that are rich in predators, usually other cichlids of the same species. The predation pressure is so high that one guardian alone could not fully protect the brood. Miles Keenleyside demonstrated the necessity for biparental care in cichlids by forcefully removing one member from a number of parental pairs and then measuring offspring survival in the presence and absence of potential predators. The parental species he studied was the rainbow cichlid *Herotilapia multispinosa*. The predators, introduced soon after the parents spawned, were either two adult male convict cichlids, two or four juvenile Managua cichlids, or two juvenile rainbow cichlids.

Fry survival was measured as the ratio of young still present 15 days after spawning relative to the number of eggs laid – the nest had been raised out of the water and the eggs had been photographed and later counted on an enlarged print. In the absence of predators (the control treatment), fry survival was around 50% in parental pairs as well as lone parents (survival was not 100% because many young commonly die owing to developmental abnormalities and within-brood competition). In the presence of predators, fry survival decreased to 30% even when both parents were present, showing that the chosen predators were not incompetents. The clincher here is that fry fared even worse when one of the parents was removed: survival was 15% when only the mother was there, and 10% when only the father remained. Obviously, biparental care afforded a much better protection to the young rainbows than the uniparental alternatives.

Helpers at the Nest

At least one cichlid species takes the concept of co-operative young-raising to extremes. In *Lamprologus brichardi* (=*Neolamprologus pulcher*), a substrate-brooder of Lake Tanganyika in Africa, offspring from previous broods stay with their parents and help them raise new broods. On average, 7-8 of these helpers can be found within the boundaries of the same natal territory, sharing all parental duties such as repelling territory intruders, removing debris, and fanning the eggs, all for the benefit of their younger brothers and sisters. As they grow older and bigger, helpers have the option of moving out to initiate their own breeding venture (often a dicey proposition because safe spots are few and far in between in their natural habitat). Some on the other hand prefer to stay a while longer in the hope of inheriting their parents' territory should one of these parents disappear. Beyond a certain body size however, helpers are often expelled by the breeding pair. Helpers try to resist eviction by adopting submissive postures, but most parents are not easily dissuaded by such pleas.

Michael Taborsky, working at the Max-Planck-Institut für Verhaltensphysiologie in Germany, conducted a number of interesting experiments with this system, both in the lab and in the field. In the lab, he observed that parents who had expelled large helpers welcomed them back into the fold when potential territory usurpers were introduced into the tank. This shows that parents value the defensive contribution of their largest (and most aggressive) helpers. Taborsky also observed that natal territories in dense tanks – where competition for space was intense – held larger helpers, which suggests that parents delayed eviction of their largest helpers when this suited their needs. In the field, Taborsky showed that helpers pay a price for their good deeds: they grow more slowly than care-free non-helpers. On the other hand, they enjoy a lower mortality rate thanks to the fact that they reside inside a territory with good shelter.

Taborsky also confirmed that natal territory inheritance takes place. In 40% of cases where he experimentally removed one member of a breeding pair, the replacement was drawn from the pool of on-site helpers. Often the replacement was the helper most similar in size to the missing parent. Taborsky and his co-workers also observed that large male helpers sometimes steal fertilizations from their bosses.

Male versus Female Care

Not all cichlids are biparental or co-operative breeders. In most mouthbrooders (those species that carry eggs and fry within their mouth), only one parent provides care. Only one parent is needed because the brood is mobile. When danger lurks, the parent can flee and take the brood with him or her. In most mouthbrooding cichlids, it so happens that the mother is the usual caregiver (there are very few instances of biparental and male-only mouthbrooders 11). This preponderance of female-only care in mouthbrooders is a mystery. When all bony fishes are considered as a group, female-only care is the rarest form of parental involvement. The most common form is male-only care. This is not very well exemplified by cichlids. For better examples we must turn to sticklebacks, sunfishes, blennies, cottids, and pomacentrids, to name but a few. In such species, males aggressively stake out territories, prepare a spawning site, and court passing females. Females who accept to spawn with a male do not linger afterwards. While the males are busy releasing milt over the eggs, the females promptly take off, leaving the males to keep on defending his territory along with the eggs he has just fertilized. The males usually tend to the eggs only, although in a few cases some attention may be granted to newly-hatched fry for a day or two.

Costs and Benefits of Parental Care

Be they male or female, alone or part of a team, most parents seem to incur tangible costs because of their dedication. Substantial weight loss during the parental phase has been documented for many species. At least one of these studies even found a correlation between the size of the brood and the extent of weight loss, large broods being more detrimental. Weight loss is probably an inescapable consequence of all the predator-chasing the parents do, as well as a lack of foraging time (or, in the case of mouthbrooders, a plain and simple inability to take in food; mouthbrooders cannot eat with their mouth full!) Demanding parental duties may so deplete energy reserves in parents that they cannot breed again for quite a while. Parents can also be stressed (what a surprise!), and sometimes it doesn't take much.

Take sticklebacks for example, a species where roving gangs of females can attack the nests of parental males with the intent of eating all the eggs inside. In Gerry FitzGerald's lab, Michelle de Fraipont kept individual male sticklebacks in aquaria where they could collect the necessary material to build a nest, and where they had periodical access to gravid females for spawning. However, the living space of all of these males was limited to only half of the aquarium. The other half was barred by a transparent partition pierced with small holes. On the other side of this partition there lived, depending on the experimental condition, a solitary female, a solitary male, a solitary female from another stickleback species, or no other fish. De Fraipont observed that only one of these experimental treatments made any difference for the physical condition of the parental male.

When the male could see and smell a conspecific female three-spined stickleback on the other side of the partition, he lost twice as much weight over the study period (80 days), he could not pack in as many reproductive cycles into this period, and in some cases he did not live as long. De Fraipont and her co-workers called this the "femme fatale effect": the stress imposed by a known egg predator, a female conspecific in this case, was said to be detrimental to the male's state of health. So there are hardships associated with parental care, but parents do get their just reward. This reward is paid in the main currency recognised by natural selection: a sizeable number of viable offspring. Parental species often breed in difficult habitats that impose a need for special care if any reproductive success is to be achieved. We saw an example above with the necessity for biparental care against predators in cichlids.

Other factors such as harsh environmental conditions may also make parental attention mandatory. One lab study with sticklebacks witnessed the progressive demise of egg clutches following the removal of their parental caretaker. Spread of disease, accumulation of debris, and lack of oxygenation were to blame for the eggs' death. Even a mere reduction in fanning levels, such as when an overabundance of territorial rivals diverts a father's attention, can have a measurable negative effect: eggs take longer to hatch, which means they remain vulnerable to predators for a longer period of time.

Egg Care: Moving the Eggs

Some substrate-brooding New World cichlids from the genus *Aequidens* lay their eggs on a submerged loose leaf. When disturbed, they move the leaf away by grabbing it with their mouth and swimming backwards with it. Miles Keenleyside and Cameron Prince conducted some interesting experiments with *Aequidens paraguayensis* (=*Bujurquina vittata*). They offered spawning pairs a choice between various artificial leaves made of black polyethylene. The leaves could be small or large (19 vs. 79 cm^2), and light or heavy (0.9 vs. 8.1 g, depending on the presence or absence of a piece of lead attached to one side). Almost all pairs preferred to spawn on small and light leaves. With a hydraulic flume, Keenleyside and Cameron proved that such leaves produce less drag.

These results are consistent with the idea that the parents choose leaves that can be moved easily. Next, Keenleyside and Cameron let a number of *A. paraguayensis* pairs spawn on leaves within aquaria with an uneven gravel surface. Some areas within the tank were under 15 cm of water while others were under 30 cm. Some areas offered good cover (many plastic plants) while other sections were bare. The researchers measured how often the parents moved the leaf, and where the leaf ended up lying most of the time, in the presence and absence of a crude predator model (a minnow-shaped Rapala fishing lure, painted to look like the predatory two-spot pike cichlid). When the predator was visible, the parents moved the leaf three times as much as usual. Most of the time, the leaf was pulled to a deep area with cover. Keenleyside and Cameron thus supported the notion that leaf-brooding and leaf-moving are an adaptation to minimise predator attacks on the eggs.

Egg Care: A "Broken Wing" Display

As mentioned above, nest-raiding is an annoying habit of female three-spined sticklebacks. Shoals of females roam and sometimes fall upon the nest of a parental male, eating all the eggs inside and thoroughly devastating it. Males take a dim view of this and they have worked out a defensive ruse. When a parental male sees a menacing shoal of hungry-looking females coming his way, he often swims a short distance away from his nest and starts poking his snout into the ground. This is the same action a female would perform while raiding a nest. This display commonly fools the females into believing that a nest has been discovered. They rush to the site and start digging there too. Meanwhile, the male leaves this writhing mass of females is similar to the "broken wing" display used by ground-nesting birds to lure predators away from their nest. Another similar behaviour has been reported for the bowfin *Amia calva*. Fry follow their male parent for a while after they hatch. Apparently, when a fry predator shows up, the male sometimes moves away and thrashes about in the water as if injured, thus drawing the attraction of the predator onto himself and away from the fry.

"Cuckoos" that Parasitize the Parental Efforts of Other Fishes

One way to guarantee egg care without risking weight loss is simply to let other fish do the work for you. Sneaky copulations are one way to achieve this. Another way is "egg dumping" (also called "brood parasitism"), a behaviour whereby females deposit eggs inside the nest of other parents and let *them* take care of the brood. In fishes, egg dumping usually takes place between species. For example, many species of minnows are known to spawn in the nest of various sunfishes. While fussing over his own clutch – preventing silting over the eggs, fanning, chasing predators – the parental male sunfish unwittingly provides care for the minnow eggs hidden within the nest. There are even some reports of a species of minnow, the golden shiner *Notemigonus crysoleucas*, dropping eggs into the nest of two of its predators, the bowfin and the largemouth bass. While doing their dirty deed, the minnows wisely stay near the tail of the bowfins, avoiding the dangerous mouth area. This system superficially resembles the nest parasitism practised by cuckoos, cowbirds, and other bird species. There are, however, important differences.

The minnows usually do not depend on the presence of host nests to breed successfully. They can spawn on their own if no sunfish nest is available (the minnows just lay their eggs in weeds and leave them to their fate without any form of care, and this seems to work well enough). Moreover, contrary to bird hosts, which have evolved numerous countertactics to foil parasitic attempts, sunfishes seldom try to stop minnows from spawning in their nests. In most cases the sunfish host does not suffer from the presence of minnow eggs within the nest. Sometimes they may even benefit. Experiments where different combinations of sunfish, minnows, and/or egg predators were mixed together and allowed to spawn showed that sunfish egg survived better when minnow eggs were also in residence, probably because of a dilution effect. When predators succeeded in piercing the sunfish's defence, the odds that they picked up sunfish eggs were reduced because of the simultaneous presence of minnow eggs.

There is one fish analogy to the true brood parasitism found in birds. In Lake Tanganyika, the catfish *Synodontis multipunctatus* attends the spawning ritual of various mouthbrooding cichlids and lays eggs at the same time, in the same spot. The female cichlid, as is her normal habit, picks up all the eggs present and commences incubating them inside her mouth. Unbeknownst to her, some of those eggs are catfish. These alien eggs hatch earlier than cichlid eggs. While still in the mouth of the female cichlid, the catfish fry finish absorbing their yolk sac. Meanwhile the cichlid eggs hatch, but only just in time to be devoured by the baby catfish! So, all that the poor female cichlid has to show for her parental effort at the end of her reproductive cycle is a few fat young that are not even her own species. One can almost picture the female catfish laughing depravedly on the sidelines. Above I wrote that egg-dumping usually takes place between species. There is one example of egg-dumping within the same species. In the peacock wrasse *Symphodus tinca*, some large males circumvent the costs of egg care by temporarily usurping the successful nests of smaller males of the same species, spawning in these nests with various females for a day or so, and then abandoning the site. The original owners, whose eggs are still present among those of the usurper, may not want to let their part of the nest contents be wasted and so they often come back and resume guarding the nest, protecting the foreign eggs as well as their own. This tactic on the part of the big males is called piracy. 27 Voluntarily caring for somebody else's eggs, and even stealing eggs to take care

In some species, females prefer to spawn in nests that already contain eggs. This preference does not leave males indifferent. In fathead minnows for example, big newly reproductive males sometimes evict the owner of an already established nest rather than occupy a similar but empty site. Such usurpers do not destroy the previous owner's eggs but instead care for them (permanently, not temporarily like the peacock wrasse above). Why do these males care for eggs that are not their own? The behaviour in fact makes sense when we learn that female fatheads, like other species, prefer to mate with males who are already caring for eggs. This preference may very well have led to the evolution of nest take-overs and adoption of eggs. In three-spined sticklebacks, territorial males have sometimes been observed dashing over to the nest of a neighbour and *stealing* eggs from him. They surreptitiously enter the nest, take a mouthful of eggs, swim back to their domain (often with the angry parent in hot pursuit) and deposit the kidnapped eggs into their own nest. This behaviour can only be explained in the light of females' preference for nests that already contain eggs. The thieving males are trying to make their nest more attractive. The Magellan plunder fish *Harpagifer bispinis* is found in shallow rubble coves along the Antarctic Peninsula. The female prepares a nest site by cleaning a patch of ground.

After her eggs are laid and fertilized, she remains on the nest, cleaning the eggs and chasing predators until hatching occurs 4 to 5 months later. This is the longest brooding period reported of any fish (everything takes longer in cold water!) Interestingly, if the female is experimentally removed, a second guardian, usually a male, takes her place. If that male is removed, a third fish, again usually male, moves in and cares for the brood. There is as yet no explanation for these altruistic acts of replacement. The new guardians may be genetically related to the original parent or to the young, but this is not supported by the fact that captive plunder fish will often accept to guard broods from other populations experimentally given to them. It must be said however that the new guardians are less diligent in their duties than the original parent, and their feeding and growth rates stay on a par with those of nonguardians, so it seems there is little cost to this type of substitute care.

Fry Care: Feeding One's Young

As mentioned above, parental care can sometimes extend beyond the egg stage, into the fry stage. In addition to cichlids, fry care can be seen in marine catfishes (family Ariidae), freshwater catfishes (Ictaluridae), in the bowfin (don't golden shiners know it), and in the damselfish *Acanthochromis polyacanthus.* Care consists mainly of protecting the fry against predators, though a few peculiarities have been documented in some species. Unlike birds, fishes generally do not feed their offspring. We can hardly expect them to feed hundreds of fry directly, nor to be able to collect and carry the microorganisms on which the fry normally feed. But there are a few particular cases. In the Kampoyo catfish *Bagrus meridionalis*, which lives in Lake Malawi, females do not lay a full complement of eggs. Some eggs are unfertilised and held back within the ovaries. When the fry are old enough (more than 15 days old), mothers gradually force out these extra eggs and the young consume them. Every day, mostly in the morning, a mother hovers 1 m above the bottom and spreads her fins slightly downwards, at which point her young raise from the nest, line up at her vent, and grab the small eggs she exudes. Kenneth McKaye has observed this behaviour and has analysed the stomach content of wild-caught young, concluding that many of these young subsist mostly on these so-called "trophic" eggs. McKaye and colleagues also reported that the father, not to be outdone, commonly ploughs into the ground near the nest to stir up debris on which the young appear to feed.

The father can also scoop up a mouthful of sediments in his mouth, churn it, and release it near the nest. The young can be seen gathering around the gills of the male as he opens and closes his opercula. They get small invertebrates that way. (Trophic eggs also exist in the cardinalfish *Apogon lineatus*, but here their role is to feed the parental male who cares for big egg broods inside his mouth, and who would be sorely tempted to eat the whole brood if he was not sustained by the trophic eggs 33. The male temporarily spits out his brood to eat the trophic eggs.) Several cichlids from the genus *Cichlasoma*, when they are accompanied by fry, are known to "fin-dig"; like the male catfish above they stir up gravel by vigorously rubbing their belly and fins against it. They also turn leaves over. Several lines of evidence suggest that this is a way for the parent to turn up food for their young, at least in the case of fin-digging. First, parents fin-dig more than non-parents. Second, parents with larger broods fin-dig more.

Third, fin-digging rate increases as the brood gets older and hungrier. And fourth, young gather up near the fin-digging parent and appear to feed actively on the stirred-up material. Of course, the possibility exists that fin-digging is not a means to provide food for the young but rather a way to obtain food for the adult itself (after all, parents, especially those that have been raising large broods for a long time, are more likely to be hungry themselves). However, Brian Wisenden and Tanya Lanfranconi-Izawa, working in the field in Costa Rica as well as in Miles Keenleyside's lab in Ontario, did not find any correlation between the frequency of fin-digging and the number of feeding bites taken by parental convict cichlids, suggesting that fin-digging was not necessarily for the benefit of the parent. On the other hand, Dmitry Zworykin, from the Russian Academy of Sciences, found that *Cichlasoma octofasciatum* parents fin-dug more when they were kept on low-food rations. Reconciling this result with the notion of fry care, he suggested that parents use their own level of hunger to estimate the need of their young. One final way for parents to feed their young also involves cichlids.

In many cichlids such as discus, Midas cichlid, angelfish, orange chromide, and others, fry can feed off the skin mucus produced by their parents. The number of visible mucus-producing glands in the parents' skin increases during the fry stage, and the young nip at the body surface of both the mother and the father. The fry do more nibbling when they are deprived of other sources of food, which shows that mucus is indeed a dietary supplement (in discus, it is in fact more than just a supplement: mucus-feeding appears to be essential for fry survival, even in the wild). 36

Fry Care: Signalling Danger to One's Young

In most parental fishes, fry care includes the signalling of danger to the young. In Siamese fighting fish for example, the parental male can communicate danger to his young through surface waves. In their first few days of independence from the airbubble nest, young fighting fish stay in contact with the surface – like most anabantids, they need to breathe some air. The parental male stays nearby and if he senses danger he shakes his pectoral fins close to the surface. The surface wavelets thus generated are perceived by the young at a distance of up to 40 cm (about a foot and a half). The young then swim in the direction of the source, and this action brings them close to the male who can then suck them up into his buccal cavity and carry them back to the nest. In cichlids, various visual displays seem to warn young in a similar fashion. A mad dashing-about by parents induces the young to settle quickly to the bottom and remain still. Brief jerks of the head or twitches of the whole body induce the fry to swarm near the parent. In mouthbrooding species, an alarmed mother can pitch slightly head-down and swim slowly backwards, upon which the fry quickly dash into her open mouth for safety.

Some cichlids, when alarmed, signal their young by flickering their pelvic fins up and down. At Illinois State University, James Cole and Jack Ward used parent models to study this signalling behaviour in orange chromides. Their models featured a pelvic fin that could be made to bob up and down by pulling on a string attached to it. They offered fry a choice between two models, one that flickered versus one that did not. By and large, the fry preferred to gather near the model that flickered. More observations by Cole and Ward, this time on intact broods with their parents, showed that parents flicker more when a small red ball is swung near their aquarium, and that fry in response form a more compact swarm. In another study on convict cichlids, Michael Shennan, Joe Waas, and Robert Lavery demonstrated that parents also flick more when they see other parents flickering. The researchers built balsa wood models of convict cichlids. They mounted these models on a cardboard background that concealed the experimenter, whose role was to manually pivot a fake pelvic fin from behind. They placed models next to the tanks of parental convict parents and either moved the pelvic fin (one flick/second for 30 seconds) or let the fin hang down. Parents reacted by themselves flickering at a high rate in the presence of the moving model; in contrast, while viewing the motionless model they flickered very little.

Fry Care: Retrieving the Young

Many cichlids retrieve their fry when danger lurks. The parents suck 2-3 young at a time into their mouth and bring them back to the old nesting site or to a safe place where they spit them out. The fry then dart or sink towards the bottom, where they stay relatively immobile. My convict cichlids do this regularly at the end of the day. This ensures that by the time darkness comes, the young are all gathered in one place over which the parents can stand guard all night long. Dimming the light at the end of the day promotes the expression of this behaviour, but mid-day dimming does not. However, complete darkness imposed at midday does trigger retrieving, the parents somehow being able to find their young and their way back to the gathering place in the dark – as I could witness using infrared lighting and infrared goggles. Here is an old anecdote reported by Konrad Lorenz, one of the founders of ethology, in his 1952 book “King Solomon’s Ring”. Late one day, Lorenz came to feed a pair of jewel cichlids he was keeping in his laboratory.

That pair had just finished retrieving their young for the night. The female was keeping watch over the pit full of fry, while the male was dashing back and forth, looking for stragglers. Lorenz dropped a piece of earthworm into the water. The female did not flinch from her guarding post but the male rushed to the worm, seized it and started chewing. Then he saw a stray fry swimming by itself away from the pit. Bent on retrieving it, he took it in his already full mouth... and then paused. What to do? To eat or not to eat? To retrieve or not to retrieve? Part of the mouth content had to go to the nest, the other to the stomach. After a few moments, the father found a solution: he spat out both the worm piece and the young. Both sank to the bottom (as I mentioned earlier, heading down is an innate response of cichlid fry being retrieved, and as for the meat, well, that was only gravity). Then the father ate the worm, taking his time and watching the nearby fry. When he was done, he took the fry in his mouth once again and brought it back to its waiting mother. Nearby students watching the scene spontaneously broke out into applause.

Fry Care: Brood Mixing

It has been observed, both in the lab and the field, that cichlid parents sometimes guard swarms of fry that are made up of several sub-groups of different body size. It seems that such parents have accepted within their brood the young from other parents (the terms "brood adoption" and "brood mixing" are sometimes used interchangeably). Indeed, it is relatively easy to experimentally integrate foreign fry into the broods of cichlids kept in aquaria. The parents accept these new fry readily, provided that the newcomers are the same size or only slightly smaller than their own young.

One of the most thorough studies of this phenomenon has been carried out by Brian Wisenden during his graduate studies in Miles Keenleyside's lab. Laudably, Wisenden worked on convict cichlids in the field (very unusual given that convicts are so easy to keep and breed in the lab). Within stretches of small streams in Northwestern Costa Rica, Wisenden captured and marked all parental convict cichlids he could find, and at regular intervals he measured the size of their broods. He observed that some broods suddenly increased in numbers while others nearby suddenly decreased, and that smaller fry seemed to have been incorporated into the new inflated broods, confirming the existence of brood-mixing and brood adoption in this population. By hand, Wisenden also transferred some fry from one brood to another. Nine times he released fry that were bigger than those from the host parents, and every time the parents *ate up* these new fry. Sixteen times he released smaller fry, and every time the parents *accepted* the newcomers.

Wisenden therefore showed that adoption is not a blind process, that parents have some say in the matter, showing a tolerance only for smaller refugees. Why should parents accept small fry but not big ones? Indeed, why should they accept any fry at all? Maybe under some circumstances they cannot tell fry apart. And then again maybe there are specific advantages to adoption. Wisenden picked up an idea that had been floating around for a long time: the host parents' young might benefit from a dilution effect when the brood is attacked by a predator. If a brood under attack was made up entirely of a parent's own young, then any success by the predator would guarantee a lower reproductive success for the parent. If on the other hand the brood was twice as big but only half of it belonged to the parent, then statistically the chances for the parent's young to be selected for attack would only be 50%. Better still, if the foreign young were smaller and predators had an easier time capturing smaller and less mobile fry, then the predator could specifically target the foreign young and leave the host fry alone. Big pay-off for the foster parents! To test this, Wisenden moved to the lab. He unleashed natural predators (three juvenile convicts or one juvenile Guapote, *Parachromis dovii*) onto broods of 20 fry, the body size of which varied within the same broods.

Looking at the survivors after 15 minutes, Wisenden saw that the smallest fry had indeed fallen prey to the predators more often than the largest ones. In view of this, why would any convict parents wish to "farm out" their young, to promote adoption of their young into a neighbouring brood? Wisenden proposed that if one of the two parents disappeared, the remaining parent might have too much trouble raising a brood on its own and might prefer to entrust the fry's fate to intact pairs nearby. Sure, the small young might suffer differential predation in their new foster family, but better that than sure death because of insufficient protection by a lone parent. Back in the field, Wisenden removed male parents from 21 pairs halfway through the fry stage (males were removed instead of females because male convicts are known to sometimes prematurely desert their family, especially if single and ready-to-breed females are abundant in the vicinity). Of the 21 uniparental broods, in 8 cases the mother steered the fry close to neighbouring groups and the fry were eventually integrated within these groups. Of the 13 mothers who decided to raise the brood on their own, only 5 saw their young survive. The other 8 disappeared, presumably at the hands of predators. 43

Longer Fry Care in the Presence of Predators

The duration of parental care at the fry stage is often dictated by the age of the young, which after a while simply become too mobile for the parent(s) to watch over. However, in mouthbrooding species, the parents have a more direct say on when to end their duties. They can simply expel the young from their mouth and refuse to take them back in. One such species has given evidence that it can extend the duration of the incubation period if it perceives that risk of predation on the fry would be high. In a laboratory experiment, females of the mouthbrooding cichlid *Ctenochromis horei* kept young in their mouth about 4 days longer (beyond a normal incubation period of 15-23 days) when they swam in the presence of another predatory cichlid (*Lamprologus callipterus*). These species are both found in Lake Tanganyika. This extra effort took a toll on the females: they could not feed during those additional 4 days, and it took them longer to breed again, as compared to females that were not exposed to the predators and that ended incubation sooner.

Filial Cannibalism

Parents who face the spectre of weight loss through the reproductive phase can counteract this effect with a rather uncaring behaviour: filial cannibalism (eating one's own eggs or fry). Eggs represent very nutritious little packages for a hungry fish. That is why egg predators are so common. Of course, parents must see their own eggs in a different light; for reproduction to make sense, hungry parents must resist the temptation to eat their own brood. Most of them succeed in doing this. But in many families (cyprinodonts, gasterosteids, centrarchids, hexagrammids, cottids, cichlds, pomacentrids, tripterygiids, blennids, belontiids), parents sometimes eat a small part of their brood, and we are not talking about diseased eggs here. Many of the consumed eggs appear perfectly viable. To test the idea that filial cannibalism is an adaptation to counter the debilitating physical effects of parental care, an experimenter needs only give supplemental food to hard-working parents.

If parents eat their eggs to avoid starvation, then cannibalism rates should decrease under conditions of plenty. Evidence of this kind has been obtained in studies on Cortez damselfish (a field study where some parents were fed eggs from other nests), a common goby (a lab study where the supplemental food was mussel meat) and the scissortail sergeant (a field study using crabmeat and eggs from other nests). Annoyingly however, even the best-fed individuals still ate a few of their own eggs in these studies. Moreover, in two other studies (one on three-spined sticklebacks fed freeze-dried shrimp and one on fantail darters fed earthworms) supplemental rations did not affect the probability nor the extent of cannibalism. Here we could object that the food supplements may have lacked some essential nutrients that could only be obtained from eggs. So, overall, supplemental food experiments leave us with a rather muddy picture about the adaptiveness of filial cannibalism. Better evidence may be gleaned from another direction: several fish observers have noted a correlation between the physical condition of the parent and its tendency to cannibalise.

In painted greenlings, river bullheads, bluegill sunfish, and the cardinal fish *Apogon doederleini*, the more emaciated a parental male is, the more of the eggs under his care he consumes. However, this does not explain why even males who are in very good condition still eat up a few eggs. A word of wisdom on this subject: there may be an adaptive side to the behaviour of filial cannibalism, as suggested by some of the results above, but we must also recognise the possibility, the one that was favoured by earlier fish observers even though it was non-adaptive, that cannibalising one's own eggs is a pathological breakdown in the normal egg-eating inhibition shown by good parents. (The current fashion in animal behaviour research is to favour adaptive explanation of behaviour over non-adaptive ones, but still the latter should not be overlooked.)

Parental Care Adjustment as a Function of Brood Value

In the late 1980s and early 1990s, the scientific literature saw a burst of publications on the topic of brood value. This was the idea that parents should be careful in how they award parental effort toward their progeny. In particular, because the business of looking after young is so costly in terms of energy, there was a risk for a parent to devote too much care to a current brood at the expense of its potential for future reproduction. If a parent was stuck with a small brood, it might consider limiting the energy invested into the care of such a low-yield evolutionary prospect, and instead save itself for better attempts in the future. All of these ideas came under the banner of "parental investment theory". Many of the published articles supported the notion that parents could adjust their level of care as a function of the value of their brood.

We saw earlier that parents sometimes eat part of their brood as an insurance against starvation. In some cases however, it is the whole brood that is devoured. This seems a bit extreme just to fend off starvation! Ethologists now consider that *total* brood cannibalism is a manifestation of parental investment theory. The parents eliminate a poor brood (which represents a low return on their parental investment) so as to be able to start a new and improved breeding attempt as soon as possible. Confirming this view, observations in many species have revealed that only *small* broods are the victim of total cannibalism. Large (and therefore more valuable) broods are left intact or only partially cannibalised. With a more experimental touch, Robert Lavery removed eggs from the nest of various pairs of convict cichlids he was keeping in Miles Keenleyside's lab.

The diminished broods were 33%, 66% or 100% (untouched control) of their former size. More pairs (6-8 out of 10) consumed what was left of their brood in the 33% and 66% treatment than in the control 100% situation (only 2 pairs out of 10). And those parents who resigned themselves to the care of reduced broods did so only halfheartedly: as compared to the controls, they performed fewer parental acts, and the ovary weight of the females turned out to be higher at the end of the experiment, indicating that they had been secretly preparing for a future reproductive attempt rather than taking good care of the current, low-yield brood. Lavery conducted another experiment on convicts. He halved or doubled some broods by transferring wrigglers or fry from one brood to another, taking advantage of the fact that parental cichlid readily accept foreign fry if those fry are of the same size as their own. He measured parental behaviour such as the percentage of time spent near the brood, the frequency of retrieving young, and the intensity of attack on a predator model (which was moved in the water through an attachment to a toy car running on a portable track resting on top of the aquaria). Lavery observed that parental behaviours were more pronounced in the brood-augmented condition than in the brood-reduced one, lending credence to the theory of parental investment. Lavery was at it again a few years later, still with convicts but this time in Patrick Colgan's lab at Queen's University in Kingston, Ontario. He measured parental response to a predator model for various broods of similar size but at different stages of development: eggs, wrigglers, and fry. He found that parents gave more and more protection to their brood as the young grew from egg to wriggler to fry. This showed that parents view younger broods as less valuable.

The rationale here is that younger broods are less valuable because there is still time in the reproductive season to initiate a new attempt, and energy reserves are still high at that time. With older broods, parents may be too weak, and the season may be too advanced to start anew. The old brood may therefore represent the only chance the parents will have, for a while anyway, to propagate their genes in the next generation, and the value of this old brood is therefore raised. Here is another variation to show that parental investment theory is a rich field for study. We are back with Brian Wisenden, still with convict cichlids. In the lab, Wisenden required his convicts to lay eggs in either a secure spawning "cave" (an overturned flowerpot with only one small triangular opening at the rim) or a risky one (an overturned flower pot again, but with two large openings). The risk stemmed from an inability to defend both openings simultaneously against egg predators. Wisenden counted the number of eggs laid in each type of cave by various females, and he found that fewer eggs were entrusted to the protection of risky caves.

It was as if the females knew that the nest was less secure, sensed that the eggs ran a higher risk of perishing, and did not dare invest too many eggs in this risky venture, preferring perhaps to keep some energy in reserve (eggs can be reabsorbed) for a future attempt in a hopefully more secure site the next time around. Ron Coleman is also a long-time student of parental investment theory in fishes, having conducted several tests with bluegill sunfish. Switching to convict cichlids, he and Alison Galvani went back to the idea that smaller broods hold less value in the eyes of their parents, and asked: well, what kind of parents? Would size of the parents matter? Would a small brood have the same value for a small parent as it does for a larger parent? For a small female that cannot lay more than 200 eggs at the best of time, a brood reduced to 100 can still have a fair amount of residual value, as compared to a large female who can lay as many as 500 eggs. Galvani and Coleman uniformly reduced the egg batches of small and large female convict cichlids down to 100 by scraping eggs off the flowerpot on which they had been laid.

This represented a relatively more extensive reduction for the larger females. Six days later, at the fry stage, the model of a tiger tilapia was moved through the tank, and the number of bites directed at it by the female was recorded. As compared to their larger counterparts, smaller females were fiercer and bit the predator model more, even though the predator probably appeared larger to them. So, brood value is a relative thing; it is in the eye of the beholder.

Female convict cichlids of different sizes do not value the same brood number equally. For a small female, investing in a small brood is not such a bad thing after all. In species where cuckolder males are present in the population, there is a chance that some of the young under a male's care might not be his. In such a case, what may matter most for parental adjustment is not brood size, but rather "effective" brood size, that is, the percentage of the brood made up of the male's own progeny. One may therefore predict that males would give less care to their brood if they can perceive that they have been cuckolded. Bluegill sunfish can provide a test of this idea. Males of this species fan and defend their eggs until they hatch (2-3 days) and protect the fry from predators until they leave the nest (5-7 days). Care is costly: the parental males do not feed and they lose about 10 % of their body weight. Unfortunately for them, other small males are cuckolders and steal fertilizations.

Parents can estimate the risk of cuckoldry from simply seeing small males in the vicinity of their nest on spawning day, and they can also estimate how much of their brood at the fry stage is made up of illegitimate offspring from their different smell. Research by Brian Neff has shown that parents can indeed adjust their level of care according to their perceived paternity. Working in the field, Neff selected the nests of males that were about to spawn, and he surrounded those nests with 4 bottles that each contained a small male. This simulated a risk of being cuckolded. Control nests were surrounded by empty bottles. The bottles were removed at the end of the spawning day, and the next day the nest owners' parental fervour was tested by pushing towards them a bottle containing an egg and fry predator (a pumpkinseed sunfish). Neff observed that the males which had been exposed to potential cuckolders directed fewer displays and delivered fewer bites to the predator as compared to the controls, in line with what the theory predicted.

He repeated the predator presentation at the fry stage, and now the treatment males defended their brood just as much as the controls, presumably because the parents could now see (well, smell) that they had not in fact been cuckolded and that the whole brood was theirs. In another part of the bluegill colonies, Neff swapped eggs (about a third of each brood) between nests. He predicted that the level of parental care would remain the same at the egg stage (since bluegill cannot distinguish between own and foreign eggs) but that the artificially cuckolded males would be less parental at the fry stage, now that they could recognise the foreign young from their different odour. This is indeed what he observed, once again based on the response of males to a predator presentation. Something similar happens in *Telmatherina sarasinorum*, a small fish found in Lake Matano in Indonesia. Increased risk of cuckoldry leads to increased rates of brood cannibalism by the father. It must be said, however, that this species is not parental. The eggs, when eaten, meet that fate soon after mating takes place. 58

Aggression in Fishes

Competition is a fact of life. It can take many forms, but biologists usually recognise two broad categories. In the first one, called exploitative or scramble competition, the contests are like races. The most food goes to the animal that eats the fastest, the best shelter is occupied by whoever reaches it first, and the largest share of eggs are fertilized by those males which produce the most sperm. There is usually little aggression displayed in such cases. However, in the second category, which is called interference or defence competition, animals fight among themselves for the right to monopolize food, to occupy alone a shelter or a territory, or to secure exclusive access to a mate. Following are some concepts and examples dealing with interference competition in fishes.

Dominance Hierarchies

Aggression allows some social fishes to sort out their relative ranks within a dominance hierarchy. Thus, when a few individuals from a social yet slightly aggressive species are placed together for the first time into a tank, a lot of nipping and chasing commonly occurs. After a while however, this aggression subsides. A pecking order has developed, every individual having figured out its place in the hierarchy. Researchers can determine the ranking of each fish by carefully observing the outcome of the initial skirmishes. The more bites an individual delivers, the more chases it initiates, and the more adversaries it wins against, then the more dominant it is. Often a linear hierarchy emerges, going from the so-called "alpha" fish at the top, to "beta" and "gamma" just below, and so on down the Greek alphabet until we reach poor "omega" at the bottom of the heap. Such a phenomenon can be observed in many salmonids, poeciliids, and centrarchids. Alternatively, the hierarchy can be despotic rather than linear. In such a case a single individual, the despot, is dominant over the other fish, who are all equally miserable. Captive eels and catfishes sometimes show this pattern.

Development of a stable and peaceful dominance hierarchy benefits everyone because fighting is energetically costly, potentially injurious, and therefore not to be done on a regular basis. However, it goes without saying that the low-ranking subordinates are not necessarily living the happiest existence. Typically their access to food is limited, as suggested by the fact that their growth rate is slower than that of dominants. For the experimenter, the challenge here resides in showing that poor growth is indeed caused directly by interference from bossy dominants, rather than poor growth and subordinate status both being caused by a third factor, such as inefficient physiology. One way to obtain such a proof is to directly observe dominants chasing subordinates away from the best food sources, as has been done with cichlids, salmon, medaka, and brown bullhead.

One can also compare the growth rate of fish raised alone and raised in groups. The usual result here is that fish raised alone show a very consistent growth rate (little variation from one lone individual to another, indicating that their physiology is uniformly fine) whereas groups yield some fish that grow well and others that do not. The slow growers usually turn out to be those that were subordinate during behavioural interactions. Moreover " and this is the clinching experimental demonstration " the growth rate of these subordinates speeds up after dominants and other competitors are either removed from the tank or isolated behind a partition. Such a sudden improvement in growth rate would not happen if weak physiology was the original cause of slow growth. Another sore point in the life of subordinate fish is that they seem to be more stressed. Stress probably develops because of the fear of being chased and chastised by bullies, and from having to worry more about where the next meal is going to come from. Stress reactions are often characterised by a rise in the production of certain hormones and metabolic products, and indeed the blood of subordinate fish often contains higher amounts of these substances. The impact of stress on the life of a fish can be substantial. In some cases, it could account for at least part of the slow growth rate of subordinates.

In an experiment conducted with Larry Dill at Simon Fraser University in Vancouver, Jeremy Abbott kept pairs of rainbow trout in tanks, each pair being made up of one dominant and one subordinate individual. These fish were of very similar size but this similarity did not prevent the dominant trout from nipping and charging at the subordinate. Now, at dinner time each day, Abbott separated the fish and fed them the same number of individual brine shrimp, fruit fly, or onion fly, which were all eaten by each fish. Despite the fact that food intake was the same for both dominant and subordinate, the subordinate ended up becoming smaller than the dominant in 10 out of 12 pairs. Abbott and Dill proposed that stress, along with the necessary investment in energy to repair fin damage, was the best way to explain why subordinates grew less than dominants despite their equal food intake. Subordinates may also be forcefully relegated to less suitable habitat. In stream species such as the mottled sculpin *Cottus bairdii* and the longnose dace *Rhinichthys cataractae*, adults often can be found in deep pools while juveniles loiter in the shallows.

It is tempting to conclude that the adults expel the less competitive juveniles from their preferred habitat and force them to eke out a dangerous living in the shallows, at risk from bird predation. However, we must first disprove the alternative hypothesis that juveniles stay in the shallows of their own volition, perhaps because their food requirement is different from the adults' and can best be met by foraging around weeds. Experimenters can tackle this problem by building enclosures encompassing both types of habitat within a stream, by placing adults in some of these enclosures but not in others, and finally by releasing juveniles in all enclosures.

If juveniles like the shallows, they should live there irrespective of the adults' presence or absence in the pools. The results, however, are that in the absence of adults the juveniles set up shop in the deep pools. With the adults present, the juveniles end up in the shallows. So there does seem to be competitive exclusion by the more dominant adults. And now for the ultimate evolutionary slight: the sex life of subordinates also suffers. In groups of guppies and swordtails for example, some alpha males have been reported to monopolise the area of the aquarium where females hang out, and to account for more than 80% of all copulations with them. This must leave the subordinate males fairly frustrated. Fortunately, at least in the case of guppies, subordinates can achieve some mating success by other means.

They can become suave rather than strong. If they happen to have brighter body colours than dominants, or if they display more, they can garner their fair share of the mating market in spite of the dominants' attempts to suppress their sexual activity. If subordinates remain within groups despite all of the inconveniences they must put up with, then probably the general advantages of social living – mutual defence against predators, discovery of food in greater amount than can be monopolised by a single alpha fish – outweigh the disadvantages of a limited existence next to bossy dominants. In fact, the necessity for harmonious group living may set a ceiling to the levels of aggression expressed by some fishes. A study by Anne Magurran and Benoni Seghers supports this idea.

In the streams of Trinidad, some guppy populations are exposed to predators and therefore show strong shoaling behaviour, while others see few predators and shoal only loosely. When Magurran and Seghers brought these fish into the lab and placed them next to a small food patch, the strong shoalers showed little aggression among themselves while eating, whereas the loose shoalers fought with one another for better position over the food patch. Cohesive shoaling and high aggression don't seem to go hand in hand very well. How stable are dominance relationships? The answer is: it depends on the species. Rainbow trout, for example, seem to remember their place in a hierarchy for a long time. Consider the following experiment by Jeremy Abbott and his co-workers. Rainbow trouts were paired up and left to establish a dominance relationship between themselves.

If one fish was 5 % larger than the other, it always won. Then, the researchers separated the two fish of each pair and fed the subordinate in excess, so much so that it eventually became at least 15% bigger than the dominant. The two fish were reunited, and surprisingly the subordinate still cowered in front of the dominant in spite of its newly acquired size advantage. Abbott and his colleagues concluded that the risk of injury during fighting is so pronounced that trout prefer to use memory rather than renewed combat to settle contests between themselves. On the other hand, catfish may constantly monitor the well-being of their competitors, looking for sudden weaknesses and chances to climb up the social ladder. In one experiment, John Todd, his graduate advisor John Bardach, and neurologist Jelle Atema, all at the University of Michigan in Ann Arbor, forced two yellow bullheads *Ameiurus natalis* to share a 190-liter aquarium. One of the two fish was clearly dominant over the other, forcing it to flee at every encounter.

When this dominant was removed, kept in a separate aquarium overnight, and then returned, the submissive recognised it right away and fled from it again. However, if during its overnight leave of absence the dominant bullhead was exposed to, and beaten by, an even more dominant catfish, then upon its return to the home tank the submissive individual *attacked* it. The defeat during the overnight contest altered something in the former dominant (its smell, perhaps, or its bravado) and this change was immediately perceived by the former submissive, who seemed to take advantage of this weakness to stage a coup. In the cichlid *Astatotilapia (Haplochromis) burtoni*, only dominant males get to occupy breeding territories.

They are surrounded by younger subordinate males who often try to take over the dominants' territories. At night, all males are pale, but at first light in the morning, dominant males brighten their blue and yellow colours, and a black stripe appears near their eyes. The subordinates stay pale. However, if a dominant disappears overnight (dip-netted by an experimenter, as it happens), then within 1 h past first light a subordinate will develop the livery of a dominant male and start courting nearby females. This shows that social hierarchies are very dynamic in this species and that social opportunities can be quickly seized upon. 11

Territoriality

A territory can be defined as "any defended area". In fishes, territories are usually held by single individuals or by breeding pairs. The defended resource may be food, shelter, a sexual partner, spawning sites, or offspring. The defenders aimed their aggression mostly at conspecifics, although other species with similar ecological requirements (or a taste for eggs and young fish) can also be targeted. Some territorial fishes are good at discriminating between full and partial competitors. They seem to realise the degree of overlap between their own requirements and those of other species. For example, in fishes that defend food territories on coral reefs, the more an intruding species shares the diet of the territory owner, the more often it will be attacked. Moreover, it will be challenged from a greater distance. In one typical experiment, individual fish from various species were placed in bottles.

These bottles were pushed incrementally towards the shelter hole of a threespot damselfish, near the centre of its territory. The diver who pushed the bottle always retreated some distance and observed whether the bottle was attacked by the resident damsel. The results: bottled fish that were known to have a strong diet overlap with the damselfish were attacked at a distance of about 1 m from the shelter hole, whereas bottled fish that did not share the damsel's diet were either overlooked or only attacked when the bottle came within less than 0.5 m from the shelter hole. Species with intermediate levels of diet overlap were attacked at intermediate distances. Good discrimination can also be found in black-belt cichlids, *Cichlasoma maculicauda,* defending their breeding territory. When there are eggs in the territory, the parents will attack egg predators and non-egg predators, but they will tolerate closer approaches by the non-egg predators before launching an attack. Moreover, egg predators that hover and seem to ogle the nest are attacked more assiduously (from a greater distance) than egg predators that just pass by.

Finally, when they have fry, parents chase predators only when these predators reach the distance from which they normally lunge at the fry. This distance varies from species to species and the parents seem to know this, for they adjust their attack distance accordingly. Territoriality is a viable strategy when two conditions are met: (1) the defended resource is sufficiently localised so that it is physically possible to defend it, and (2) there is some competition for this resource, but not too much. Let's illustrate both principles by using food as an example of defended resource. First, the question of localised resource. Suppose a fish needs a certain daily amount of food to survive. If this amount can be found in a relatively small area, then this food clump is relatively easy to defend and territoriality becomes a suitable strategy.

If the food is more spread out, territoriality can still take place but the territory will have to expand " and therefore become harder to defend. At the extreme, if food is very thinly distributed, the required territory size would have to be so large that it would be physically impossible to guard all of its boundaries. In such cases, territorial behaviour is absent. These ideas lead to testable hypotheses. One is that territory size should increase as food becomes less abundant, but only up to a certain extent. Territory size can be estimated by linking, on a map, all of the outermost places where intruders are attacked by a given defender. It is then possible to measure the size of the area thus delimited. Food abundance, for its part, can be sampled by researchers within the territory itself (much to the dismay of the territorial fish who usually cannot defend against big human intruders " although some puny but pugnacious damselfishes have been known to try). In this way, salmon and trout have been shown to indeed occupy larger territories when food is less abundant. With a more experimental touch, Mark Hixon, of the University of California at Santa Barbara, found territorial male black surfperch *Embiotoca jacksoni*, a coral-dwelling species, and decreased the amount of food accessible to them. He did this by covering portions of the coral reef with swatches of nylon netting through which the fish could not pass.

Unfazed, the surfperch reacted by increasing the size of their territories and appropriating neighbouring patches of untouched coral. However, when the reduction in food supply was very great, as occurred when overgrazing sea urchins invaded the area, the surfperch abandoned their territories rather than try to expand their boundaries. The disappearance of territorial behaviour when food becomes too thinly distributed has also been documented in controlled lab experiments. In most cases, automatic feeders were set up over a very large aquarium containing many fish. The feeders were programmed so that only one of them dispensed all the food, or each one of them gave a little bit of food. The prediction was that fish would become territorial in the first instance but not in the second, and this is indeed what happened. When only one feeder was activated, the dominant fish in a group soon started to defend the area around that feeder. When all the feeders were activated, it became impossible to defend all of them.

The group therefore spread out and swam all over the tank. Very few individuals, if any at all, bothered to defend a given area. This was observed in studies with medaka, pygmy sunfish, salmon, and juvenile cichlids. In the wild, feeding territories are seldom observed in freshwater habitats, but they are fairly ubiquitous over coral reefs. One possible explanation for this state of affairs is that coral reefs offer a richer supply of food " their rate of production has been estimated to be 10 times as high as that of an average lake or stream. Therefore, over coral reefs, food can be concentrated in one area sufficiently small to be successfully defended. Conversely, the relative lack of territoriality in freshwater habitats could be linked to the greater dispersion of food sources. This is not to say that freshwater species are incapable of establishing territories when conditions are right, in other words when resources suddenly become clumped. The species mentioned in the preceding paragraph, where dominant individuals defended territories around single feeders in the lab, were all freshwater species. Now, on to the question of competition level. If food is superabundant everywhere, so much so that everyone gets to eat to their heart's content and no competition exists, then obviously there is no point in establishing a territory. At the other extreme, if the number of competitors and intruders is so high that a single fish cannot defend even the smallest of territories against everyone, then obviously there is no point in territoriality. It is at intermediate levels of competition that territoriality becomes a viable prospect.

These ideas can also be tested in the lab. One can set up a great number of feeders over a tank, and switch from a situation where only a few feeders are giving food (territories will probably be defended around them) to an all-you-can-eat buffet where all the feeders are offering a lot of food. In such a tank of plenty, harmony should reign supreme and aggression should subside. In medaka, this is what happens. Interestingly, if some of the medaka get paranoiac and persist in their belligerent ways even though food is superabundant, they turn out to grow more slowly, probably because of all the energy they needlessly spend in chases and threats directed at others. Intruder pressure, for its part, can be experimentally intensified simply by adding fish to a tank.

A number of species have thus been shown to have a harder time maintaining a territory when competitor density is very high. One of the first reactions of territory owners when intruders become too numerous is to decrease the size of their territory to make it easier to defend against this onslaught of trespassers. Later, if competitors (or alternately, territory neighbours) are experimentally removed, the remaining residents expand their domain back to what they see as an ideal size. Note that topography can also influence territory size. In habitats that are structurally complex, with lots of rocks or plants (some of which may have been added by curious experimenters), territorial defence is more difficult because boundaries cannot be visually monitored all at once. This results in smaller territories, and in more fish cohabitating next to one another. Topography may also provide natural landmarks that act as bastions for territorial defence, and if such landmarks are abundant, the tendency to use them as borders may lead to smaller territories.

What Makes a Good Fighter?

If a fight erupts between two fish, can we predict which one will be the winner? The answer is yes if there is a big size difference between the contestants. As one might expect, big fish have the upper hand. One example of this comes from Indiana University, where William Rowland kept a large number of three-spined sticklebacks in stock tanks. These fish were somewhat crowded in a bare environment and could not establish territories. Rowland dipnetted various males from the stock tanks and weighed them. To put these males in a fighting mood, he placed them into individual aquaria and let them establish breeding territories. Then he picked two of these territorial males at random and moved them together to yet another aquarium, one that was unfamiliar to both of them, although it still looked like their own. The two males soon faced each other and initiated a fight.

After much spine-erecting, head-to-tail chasing, and biting, the loser declared itself by breaking off the fight and cowering in a corner. Rowland staged 31 such encounters in which a clear winner emerged, and he found that the heaviest male was victorious in 22, or 71%, of them. Statistical methods revealed that the greater the weight difference was between contestants, the greater the chances that the heavier fish would end up winning. A weight difference of 15% practically guaranteed victory for the heavies.22 In the experiment above, it was important to stage the contest in a neutral arena unfamiliar to both fish. If the contest had taken place in the home tank of one of the two sticklebacks, the territory owner would have held a much greater probability of winning than might be inferred simply from its body size. Being in one's own territory seems to confer more confidence, or perhaps a greater realisation of what is at stake for the owner. Sports fans call this the home turf advantage.

Ethologists prefer to speak of a "prior residency effect". Within certain limits, the prior residency effect is enough to prevent large intruders from usurping the territory of smaller residents. The duration of prior residency may also have an influence: in brown trout *Salmo trutta*, longer-term residents (4 days) outperform shorter-term residents (2 days) in the defence of their territory. Another factor that may influence the outcome of a fight between two closely-matched fish is a prior experience of submissiveness. A fish that has just lost a contest is more likely to give up during the next fight as well, even when this second fight is against a new adversary. It is as if the first setback created a general losing state of mind. In sticklebacks, this lingering "loser effect" can last up to 6 h. As an example, we can look at the work of Theo Bakker and his colleagues at the University of Leiden in the Netherlands.

These researchers kept a great number of territorial male sticklebacks in individual tanks. They chose a few individuals at random and subjected them to a losing experience by dropping them into the tank of another male: this other male beat them up, taking advantage of the prior residency effect since he was in his home territory. Other males were also chosen at random and they experienced a win by having another male dropped into their own tank (for them, the prior residency effect worked in their favour). Three or six hours later, these respective losers and winners each met an inexperienced male within the confines of a neutral arena unfamiliar to all of them. If it was not for their previous experience, we would expect the previous losers as well as the previous winners to dominate this new encounter on only half of all tests since all of these fish were chosen at random. Previous winners indeed won only half of the time, indicating that their previous winning experience did not make them stronger.

But a different picture emerged in the case of the previous losers: none of them won a single fight when this fight was held 3 h past their first debacle. Even after 6 h, the previous losers won on only 20% of all tests. Similar results have been obtained with blue gourami, paradise fish, green swordtail, and pumpkinseed sunfish. In the case of the blue gourami and the pumpkinseed sunfish, it seems that a previous *winning* experience can instill confidence and help a combatant win its next encounter. This winner effect does not last very long in the sunfish – no more than 1 h – but it does persist for at least 3 days in the gourami. There is also the Mangrove rivulus *Rivulus marmoratus*, where both a winner and a loser effect exist for about 2 days. Maybe prior experience fine-tunes the information a fish has about its own fighting ability. Or maybe the fight alters hormone production differently in winners and losers, with an impact on their willingness to fight again. (This hormonal hypothesis has been invoked to explain another short-term winner-loser effect: in the Mozambique tilapia *Oreochromis mossambicus*, 15 minutes after a fight, winners court females more readily, for a longer time, and with more courtship sounds than losers, even when each male is alone with the female.) Fighters are more successful when they can first impress their opponents with signs of their good health and good growth. Sometimes, these signs are the same ones that are used to woo females.

In swordtails for example, males possess an elongated lower tail section that resembles a sword. Females prefer to mate with males who have longer tails; interestingly, males with longer tails also win more fights, even when matched with opponents of similar body size. It is possible to attach plastic extensions to the tail of a given male, and all of a sudden this male starts to win more fights than he used to do, probably because his big sword incites his adversaries to back down. In the same vein, male sticklebacks with brighter red throats attract more females, and also win more fights against other males. Intimidation is probably involved rather than actual fighting ability, because when fights are staged under blue light (which makes the throat appear black instead of red), males with brighter red throats do not win more fights any more. Intimidation is behind all of the ritualized displays that are performed by both contestants at the beginning of a fight. In fishes, such displays include booming sounds, water-displacing tail beats, fin erection, gill cover spreads, head shakes, body twists, lateral displays that reveal the full size of the body, colour changes, exposure of brightly coloured body parts, and intricate swimming manoeuvres.

These actions are meant to signal fighting ability and to encourage opponents to give up. Other determinants of dominance during fights include stamina and motivation. Escalated fights may last for a long time (a half-hour is not uncommon in some species) and stamina would prove an asset in such a situation. Consider the work of Francis Neat and co-workers on the redbelly tilapia *Tilapia zillii*. These researchers found that losers of territorial fights harboured more lactate within their muscle than winners did immediately following the fight. Lactate is a metabolic by-product that can cause fatigue. So the vanquished fish may have lost because they were the first ones to get tired out. Motivation, or "fighting spirit", could also characterise good combatants. In another study on redbelly tilapia by Neat, small males sometimes won over larger ones. These smaller winners were more aggressive during the fights and inflicted more bites. They also had larger gonads, indicating that they were more ready to spawn, and therefore perhaps more inclined to defend their breeding territory. Being closer to spawning may also explain why breeding pairs of convict cichlids that have been together for a longer rime (96 h versus 48 h) fight more successfully for breeding sites. In some species, motivation to fight may also be influenced by a "priming" effect.

If a male fighting fish who has just witnessed a combat between two other males is allowed to interact with a male who has just seen two other fish not fighting, the former usually behaves more aggressively than the latter. It's as if viewing a fight put the male in a fighting mood. In the same vein, three-spot gouramis who have learned to associate the appearance of a red light with the imminent arrival of an opponent win more fights when they are forewarned by the red light.

How Fishes Try to Avoid Predators

Fishes are delicious. At least herons, kingfishers, mergansers, and all marine birds seem to think so. Marine mammals seem to think so. Minks and bears seem to think so. The big fishes themselves concur. As a taxonomic group, fishes face one of the most diverse arrays of predators imaginable. Threat comes from below and from above, during the day and at night, and at almost all stages of life. Is it any wonder that most fishes – certainly all the small ones – are skittish creatures? When placed in a new environment, most wild fishes cower in nooks and crannies and don't dare raise a fin. They don't want to draw the attention of predators. But of course, an animal cannot spend its whole life immobile. Fish may remain in constant fear of danger, but they also have other things to do, feeding and reproductive activities being the most obvious. The life of fishes is dictated by a triumvirate of imperatives: the need to reproduce, the need to eat, and the need to avoid being eaten. The problem is that these activities are not always compatible. Compromises must be achieved between courting a potential mate and looking out for enemies. Trade-offs must be accommodated between the necessity to search for food and the desire to avoid detection by a predator. This looks at the nature of these compromises, with predation risk as the underlying causal factor. There has been extensive research on this topic.

This reflects not only the importance of predation risk in the life of fishes, but also the relative ease of experimental manipulation. The basic protocol has been to compare the behaviour of a fish before and after the appearance of a predator. The comparison can also take place between a group of fish that sees a predator versus one that does not, or between natural populations that experience different levels of predation pressure. In the lab, predators can be presented in a multitude of ways. A predatory fish species can be positioned in plain sight within the confines of an adjacent aquarium. A tame predatory bird can be tethered nearby. Resin models of a big bad fish can be cast and then dragged through the water. Wooden models of heron heads can be thrust fiercely through the surface. Cardboard silhouettes of kingfishers can be "flown" along a wire above the water. Fish can also be exposed to an inflow of water laced with the smell of a predator or the alarm substance of its prey. The response of fishes to predation risk can be divided in four broad categories: (1) the fish can switch habitat, i.e. decamp to areas where there are fewer predators or better shelters; (2) they can stay in the same habitat but keep a low profile, reducing the frequency of conspicuous behaviours and the amount of time they spend exposed; (3) they can remain exposed but increase the percentage of time spent vigilant, usually at the expense of the time and concentration necessary to do other things; and (4) they can shift their activity to other times of day when predators are not so abundant or not so successful.

Switching Habitat

Predator-free ponds provide great opportunities for studying the effect of predation risk on habitat choice. All you need to do, with the permission of your local environmental government agency, is to introduce a predator to a site that encompasses various habitat types – open water and weeded areas, for example – and compare the spatial distribution of the prey before and after predator introduction. Alternatively, the pond can be fenced up into two halves, one with prey only and one with both prey and introduced predator. This has been done with bluegill sunfish and largemouth bass, minnows and pike, young crucian carp and Eurasian perch, and young Eurasian perch with their cannibalistic elders. In all cases, prey in the predatorfree condition occupied both open waters and shallow weeded areas, whereas prey exposed to predators stayed in the shallow weeded areas most of the time. Confinement to the shallows usually led to slower growth rates, because food was not as plentiful there, and because there was more competition for it from the great concentration of refugees. In the above experiments, some prey could be seen to venture, sometimes even to set up shop, in the open waters where the predators operated most efficiently.

Invariably these fearless individuals were large. Great size does confer some degree of immunity against predators. These large individuals were not constrained by the competitive bottleneck that affected their smaller brethren in the shallows. Therefore they grew more quickly and became even safer from danger, a case of the rich getting richer and the poor staying poor. (The anti-predator benefits of large size are further illustrated by an intriguing observation. In the absence of predators, crucian carps develop slim bodies that are hydrodynamically efficient. But in the presence of predators, carps grow to become rounder and larger, a body shape that is not so economical for swimming, but more likely to deter predators because it is not so easy to swallow. It seems that the differential growth is induced by exposure to carp skin substances exuding from the faeces of the predator.) Predation experiments can also be done in the lab. For his PhD thesis at Queen's University in Ontario, Vytenis Gotceitas built artificial weeded areas by attaching green polypropylene ropes to grids of wire mesh. Within wading pools, he installed patches of "weeds" in densities of 50, 100, 250, 500 or 1000 stems per square metre.

He also introduced a great number of damselfly nymphs, a natural food supply for bluegill sunfish. He then released bluegills into the wading pool and observed their behaviour before, during, and after the introduction of a piscivorous largemouth bass. Before the predator appeared, the bluegills stayed in open water or in low-density weed patches because this is where they had the most success finding and catching damselfly nymphs. When the predator was thrust upon the scene however, most bluegills moved to the high-density weed patches.

That was a good choice, because Gotceitas could see that the largemouth bass was fairly successful at catching those few individuals that stayed in the low-density weed patches. Therefore, dense weeds were a good place for bluegills to seek refuge from predators. Let's remember though that dense weeds were also a poor place to forage. So, in the absence of predatory activity bluegills patrolled the sparse weeds to maximise their foraging success, but they moved to safer patches of dense weeds when a predator appeared, even if that meant poor foraging, a lab result that mirrored the field observations above. If habitat switches entail a trade-off between foraging and avoiding predation, then it should be possible to experimentally manipulate this balance and tip it either in favour of more foraging despite the risk of predation, or conversely more sheltering despite the risk of starvation.

The simplest way to do this is to compare the behaviour of hungry and satiated fish. Both can be offered a choice between spending some time in a safe habitat devoid of food, or a risky one where there is food. This has been done in the lab for crucian carp facing pike, black gobies facing cod, pink salmon fry facing adult chinook, and juvenile coho salmon facing adult rainbow trout. In all cases, the hungry individuals spent more time in the risky area, close to the predator but with good access to food, than the better-fed fish. Those experiments hint at another way to affect the balance of foraging opportunity and predation risk. We can vary the quantity or the quality of food in the risky habitat. A choice can be offered between two patches, one that gives access to a little food and that is placed near an adjacent aquarium that contains no, or maybe only one, predator, versus another patch that offers more food but that is also next to an aquarium containing two predators.

The question is: how much more food should the more dangerous patch contain in order to draw the wary prey there? Experiments of this kind have been done with juvenile creek chub facing predatory adults, young black surfperch at risk from kelp bass, European minnows exposed to a kingfisher, guppies facing cichlids, and upland bullies viewing a salmon. The switch from safe to dangerous habitat took place when food was at least 3-4 times, and sometimes as much as 28 times, more abundant in the risky site, a substantial difference that may not always be present in natural situations. This could explain why, in the natural experiments described above, fish at risk from predation stayed in the shallow areas of ponds and lakes despite the lower food supply there. If fish have at least enough food to survive in the safe habitat, and the dangerous habitat is not that much better in terms of food availability, then prey may elect to stay in the safe habitat most of the time. Yet another way to tip the scale is to alter the availability of refuges in the various habitats.

Prey may accept to venture in predator-rich areas if there is also structure, there to protect them. To demonstrate this, Douglas Fraser and Richard Cerri built compartmentalised channels within a spring-fed stream in the Hudson-Mohawk River watershed. Within each compartment they could manipulate the presence or absence of a predator (adult creek chub) and the structural complexity of the habitat (pieces of black pipe, wood, covers providing shade). The compartments were separated by wood dividers with slots big enough to let small minnows go in and out but too small to let the predators exit. Small minnows (young creek chubs and blacknose dace) were let loose in those channels, free to move from compartment to compartment. Their distribution could be determined at any time by dropping hinged gates which effectively made all fish prisoners of the compartments in which they happened to be at that moment. In this way, Fraser and Cerri observed that minnows tended to avoid compartments with predators but that this avoidance was less marked when structure was present in those compartments.

Predator avoidance is a strong incentive at all times but the presence of structure can mitigate it somewhat. Similar results have been obtained in the lab with other species. Of course, habitat shifts may not afford complete safety. Some predators have this nasty habit of adapting and venturing into the areas where their prey take refuge – predators have to make a living too, you know. For example, largemouth bass can switch from cruising in open waters to ambushing in vegetated areas. Small prey fish may flee from harmful perch in open waters only to fall prey to a stalking pike in the weeds. Minnows may think they are safe from large predatory fishes in the shallows, but then they are nabbed by a heron. As I said earlier, fishes are just too tasty. They are almost never completely safe. Nevertheless, the fact remains that habitat switches can at least help to decrease predation risk. Better a small risk of being caught by a vpike in the weeds than guaranteed death from a bass in open waters.

Reducing Conspicuous Behaviours

Juvenile salmon usually hold station somewhere in a stream and occasionally dash upstream to intercept drifting prey. Larry Dill and Alex Fraser from Simon Fraser University wondered how this behaviour could be affected by predation risk. They compared the foraging behaviour of coho salmon that could feed under two different conditions, either undisturbed or after being distracted by the presentation of a photograph depicting an adult rainbow trout (a predator of young salmon). Their results were that, all other things being equal, the cohos that had seen the photograph were not willing to swim as far away as usual in order to catch drifting insects. Whereas unperturbed salmon were willing to swim 25 cm upstream in order to catch a big fly, disturbed salmon would only go 16 cm. In Glasgow, Neil Metcalfe and his co-workers observed a similar reticence to venture out on the part of scared Atlantic salmon, adding that it took 2 h for the feeding behaviour to fully get back to normal after predator presentation. The inference from both of these studies is that wary fish probably want to minimise the amount of time spent moving.

Other studies in Dill's lab have shown that moving salmon are attacked by common mergansers more often than stationary ones, and maybe the same would apply to predation by large trout. Here again it is possible to alter the trade-off between safety and foraging by playing with the hunger level of the salmon, or with the levels of predation risk. Dill and Fraser manipulated their cohos in this way. They observed that hungry salmon reduced their attack distance on drifting prey when scared by a predator, as expected, but not as much as better-fed individuals did. Because the cohos were hungry, they were willing to take a little bit more risk. Salmon which could see their own image in a mirror were also willing to take more risk by dashing a little further than lone individuals. Either they perceived the mirror image as a competitor for food and consequently they were more motivated to get the food, or they felt safer because they had a companion and reckoned there was less chance for them to be the specific target of an attack.

Dill and Fraser also manipulated the balance in another way: they varied the frequency with which the predator image was presented. As expected, salmon which were exposed to the image of a predator more often (every 22 minutes) reduced their attack distance to a greater degree than salmon who saw the predator less frequently (only at 45-min intervals). The fish were able to estimate the higher level of risk and adjust their foraging behaviour accordingly. The need to avoid conspicuous behaviour in the presence of a predator can also have\ an impact on a fish's sex life, especially the males' courtship behaviour. Let's take the case of guppies. Males have two ways of mating with females. They can woo them with a sigmoid display, in which the body is arched and the fins are extended. Such a display is conspicuous, can take up to 5 seconds to perform, and must be done often before a female finally agrees to mate.

The second strategy is sneakier. It is called gonopodial thrusting, a forceful insemination without the female's co-operation (a form of sexual coercion). Gonopodial thrusting is less conspicuous than sigmoid displays, but the chance of a successful insemination is also reduced because the female tries to resist it. The interesting point here is that when we compare the relative frequency of both strategies in the presence and in the absence of predators (cichlids or characids) at large in the same environment, the sneaky behaviour predominates when predators are present while the conspicuous display is more important in the predator's absence. It seems that predator-wary fish abandon effective but conspicuous courtship displays and resort to less showy but safer alternatives if they can. Another option under predation threat is to shorten the duration of courtship before finally mating, as has been observed in pipefish, sand gobies, darters, razorfish, damselfish, and sticklebacks.

Camouflage

There is one category of fishes for which reduced activity is an integral part of antipredator strategy: cryptic species whose body colour matches the surroundings. For camouflage to be effective against a static background, the fish must itself remain motionless. There is evidence that freezing in cryptic species is a conscious effort to blend in and not simply an attempt to reduce conspicuous movements irrespective of the potential for camouflage. Tidepool sculpins, whose body markings mimic the appearance of sand, have been kept in aquaria with either a matching (sandy) or nonmatching (white) bottom. When scared by the introduction of an alarm substance, the fish on matching sand reduced their movements to 65% of normal levels, as might be expected. However, the fish on a white substrate did not alter their activity. For them, immobility would have conferred no cryptic advantage, and consequently that tactic was not adopted. Active search for a refuge was a better alternative in that case. Another example comes from a study of three darter species.

The fantail, greenside and orangethroat darters wear dull colours outside of the breeding season and they freeze over mucky bottoms in response to predator signs. During the breeding season, the male fantail and greenside darters develop a conspicuous green body colour, but because they normally breed near matching green algae they keep on freezing when alarmed. In contrast, male orangethroat darters develop intense orange, blue, yellow and red breeding colours. Needless to say, they cannot find matching surroundings, and therefore it comes as no surprise that they abandon freezing as an anti-predator tactic and resort to fleeing instead.

Increasing Vigilance

If a fish is confident that it can escape from a predator as long as it has enough advance warning, then all it needs to do in a risky environment is to increase its time spent vigilant. For the fish ethologist, studying this topic poses a problem: how do you measure vigilance? How can you tell that a fish is vigilant? A fish cannot perk up its ears like a mammal. It cannot look up like a bird. We therefore have no choice but to resort to a more indirect sign. In most cases, what ends up being measured is foraging activity. This may seem completely unrelated, but the rationale is in fact sound enough: foraging requires concentration on the task at hand and is therefore incompatible with vigilance. Good vigilance demands a fish's undivided attention and it cannot be done while the fish is feeding. The intensity of foraging can therefore be construed as an inverse index of vigilance. Ethologists therefore predict that fish should reduce their feeding rate when they perceive a risk of predation. Satisfyingly, this has indeed been observed in a variety of species, most notably salmon and sticklebacks. In one experiment conducted by Manfred Milinski, three-spined sticklebacks were placed in an aquarium in which stood a number of upright Plexiglas cylinders. At the bottom of each cylinder was a tasty Tubifex worm. However, the cylinder was of such a height that the fish lost sight of their surroundings while reaching for the worm at the bottom. In one treatment the sticklebacks were on their own, while in another treatment they could see the predatory cichlid *Oreochromis mariae* through a nearby transparent partition.

The unthreatened sticklebacks fed enthusiastically in all cylinders, but the wary fish reached for the worms less often and when they did, it was only in those cylinders furthest away from the cichlid. We can infer that they needed to remain vigilant. In Scotland, Neil Metcalfe and his co-workers placed juvenile salmon in an artificial stream channel and dropped food pellets a short distance in front of them. The current carried the pellets past the salmon, who could first orient towards the pellets and then "attack" them. Two pellet types were used, one which was too large to be swallowed and one which was just the right size. The salmon had been familiarised with both types and knew the difference between them. An experimental trial consisted of dropping a total of 6 pellets, 3 large and 3 small ones, one at a time at 10-minute intervals. One group of salmon was left undisturbed, but another group was shown the Fibreglas model of a predatory brown trout for 30 seconds before the trial began.

Though brief, this presentation had an effect: during the hour-long trial that followed, the frightened salmon attacked the drifting pellets less often, and when they did they seemed not to discriminate very well, attacking the inedible large pellets as often as the edible small ones. The undisturbed controls attacked the edible pellets at a high rate and the inedible ones less often. These results suggest that fear of predation, and the consequent need for vigilance, rob juvenile salmon of the concentration needed to discriminate between food items. Concentration is also required of a fish that feeds on high-density swarms of *Daphnia*. This is because of the confusion effect; we had seen that predator confusion could benefit large shoals of prey fish, and it is not hard to imagine that it can also benefit large swarms of *Daphnia*. One experiment with guppies has confirmed that the concentration needed to overcome the confusion effect hinders vigilance and predator evasion. Guppies feeding on *Daphnia* at densities of 1, 5, 10, 15, or 20 per litre were subjected to surprise attacks by a live jewel cichlid.

The outcome for the guppies depended on the density of the *Daphnia* on which they fed: the greater the density of *Daphnia*, the greater the probability of the foraging guppies being caught by the cichlid, from around 20% at the lowest density to 50% at the highest one. Overwhelmed by the whirlwind movements of all those little prey items in front of them, guppies did not see the predator coming and paid dearly for it. Consequently, we would expect wary fish to feed on low rather than high-density swarms, since this would require less concentration and allow better vigilance. Milinski has provided evidence that this is so. He filled test tubes with various numbers of *Daphnia* (0, 2, 20, or 40) and presented all tubes simultaneously to individual sticklebacks. Some of these sticklebacks had been previously frightened by the overhead flight of a model kingfisher, while others had been left alone.

Milinski saw that most of the undisturbed fish at first attacked the tube that contained 40 *Daphnia*, probably because it represented a rich source of food and the fish, being unaware of any predator in the vicinity, were willing to invest the concentration necessary to try and isolate prey one at a time (eventually though, they switched to the less packed tubes, maybe in frustration at their initial lack of success reaching the protected Daphnia within their tubes). In contrast, most of the frightened fish first bit the tube that housed only two *Daphnia*, and they maintained that choice. They wanted to keep an eye out for the return of the predator they had seen before and preferred prey that could be caught quickly, even if that meant settling for fewer of them. Although the above examples all deal with foraging as the antithesis of vigilance, there is no reason to believe that other behaviours besides foraging could not be measured with the same intent. Reproductive activities such as mate choice, nest-building, fighting with territorial neighbours, and caring for eggs also require a certain amount of concentration and can therefore be implicated in a trade-off with vigilance. Already some researchers have reported that female guppies stop paying attention to courting males, or show less discrimination between them, when they perceive a predation risk. There is scope for more research involving such vigilance-incompatible behaviours.

Altering the Timing of Activity

With the help of a few students, undergraduate student Lyne Boudreau and I once placed minnow traps in a stream to see if lake chub would be caught mostly during the day or at night. I expected the answer to be during the day because chub kept in aquaria are almost exclusively diurnal. To my surprise the chub ended up being caught only at dawn and dusk. This was with unbaited minnow traps. When we baited the traps with dry dog food pellets, the chub were caught at dawn and dusk as before, and also during the day, as previously expected. We interpreted these findings as follows: the chub were among the largest minnows in the stream and under the full light of day they were particularly visible to kingfishers and mergansers, two fish-eating birds that had been spotted in the vicinity. Accordingly, the chub restricted their activity to dawn and dusk, a time when low light levels impaired the hunting behaviour of the birds. Only when the balance between predation risk and foraging success was tipped in favour of foraging, by adding nutritious bait to the traps, did the chub accept to venture out during the day. This example suggests that fish may shift the peak of their activities to those daily times when predators are less active or less successful. It is no strong proof however.

Maybe the chubs were crepuscular because their preferred prey happened to be crepuscular as well. A convincing experiment would require all avian predators to be removed from the vicinity of the stream, in the hope that the chub's activity would then shift back to being fully diurnal. To do so would be impossible, for practical as well as ethical reasons. The situation might be more tractable in the lab, as predators could be presented at the same time every day, day after day, in the hope of teaching the fish to reduce activity at that time and to compensate by becoming more active at other times. I am aware of only one experimental study that has convincingly linked predation regime and a prey's shift in diel timing of activity. Douglas Fraser, James Gilliam, and collaborators conducted a field study in Trinidad, in which they looked at guppy behaviour in predator-free and predator-present pools.

Hoplias malabaricus. The scientists found that guppies were strictly diurnal in the presence of the predator, but were active day and night when free from predation. Night foraging was as profitable as day foraging, and therefore the guppies grew much better in the predator-free condition. So this was a case where predation seemed to limit the activity phase of a fish who could otherwise be active all the time. This page delved into the effect of predation risk on fish behaviour. I insist on the word "risk". In all cases covered here, the fish were wary but they were not under direct attack from an enemy. When an attack does happen, the behavioural response of prey fish is fairly straightforward: they flee or try to hide. How long they remain hidden or "frozen" depends on how scared they feel, and how eager they are to resume feeding or courting. Shoals under attack can also "explode", with all fish swimming in all directions (a reaction called flash expansion), or they can show a "fountain effect", splitting up in two, each halves passing by the predator's sides before rejoining behind it. Other fishes rely on anatomical and physiological defences rather than behaviour. They grow bony plates and spines on their body, or they synthesise toxins which are stored in skin or flesh. Some develop body markings that mimic the appearance of foul-tasting species, hoping to fool experienced predators into leaving them alone. Others sport false eyespots on their tail, and this may deflect predator attack away from the sensitive head area, or confuse the predators when the prey suddenly starts swimming "backwards". It is tough being a fish. Everybody wants to make a meal out of you. It must make for a stressful existence. Yet, fishes endure. Some of them even thrive (with the notable exception, these days, of those species that are commercially-exploited by people, a smart predator with whom it is hard to cope). Fish survival in the face of so many predators bears witness to the care fishes take in minimising predation risk.

How Fishes Find their Way Around

One of the joys of keeping fishes as pets is that you don't have to worry about them being unhappy because of confinement. Many fishes have a natural preference for small home ranges and they are quite satisfied with life in an ordinary aquarium (as long as you provide some objects for them to hide or swim around). This, however, does not mean that there isn't a need for proper spatial orientation in fishes. In nature, even a fish that spends all of its adult life within the boundaries of a small home range may occasionally get displaced by storms or by predators. After such misfortunes, the fish must be able to find its way back home, where the location of good food spots and hiding places is known. Moreover, large species, which are not very suitable for the pet trade and therefore not common in aquarists' living rooms, are often quite mobile in nature and may travel over wide areas. The seasonal migrations of salmon and other species are well known in that regard. Some fishes even embark on wide-ranging travels on a daily basis. In a study conducted in the Gulf of California by Peter Klimley and Donald Nelson, scalloped hammerhead sharks, *Sphyrna lemini*, were fitted with ultrasonic transmitters that allowed their movements to be tracked via receivers on boats.

During the day the sharks were idling around a small seamount, but at dusk they left on foraging trips that took them as much as 8 km (5 miles) away, and yet at dawn they unfailingly returned to their point of departure. Even more impressive are the journeys of one skipjack tuna, *Katsuwonus pelamis*, which was tracked by similar means off the island of Kauai in the Hawaiian archipelago. This fish could go as far as 35 km (22 miles) away from its usual resting place, and yet it always came back to that spot at the end of a foraging trip. Ultrasonic telemetry is useful in studies of this kind because the route taken by the homing fish can be plotted (more precisely, the route taken by the boat that follows the fish is plotted, but that works out to be pretty much the same thing). More information is thus available to figure out how the fish finds its way. Unfortunately, the technology also has drawbacks. The transmitters are expensive, have a lifetime of no more than several weeks, and are so bulky as to restrict their use to large fishes only. For fun, compare this to the 1950s' way of tracking fishes, a method that was low-tech and that still worked well for species living in lakes. Hooks were simply inserted through the dorsal musculature of a fish and attached to a small float via a long thread.

The movement of the float at the surface would betray the movement of the fish below, and could easily be charted. Imagine the surprise of boaters unaware of the ongoing study when they saw a piece of Styrofoam moving erratically on the surface of the lake! More popular than telemetry or Styrofoam-tracking is the method of capture displacement- recapture. In essence, this method consists of catching a fish in its home range (the area of which has already been determined by direct observations or by previous captures), tagging the fish, and then releasing it at a remote site. Traps are then set on the original home range, as well as elsewhere in the habitat, and this is done for many different individuals. If the displaced fish can find their way around, each one of them should get recaptured in the home traps and not elsewhere. However, if the fish cannot orient properly, recaptures should follow a random pattern in all traps.

Through experiments of this kind, some impressive feats of homing have been documented. In New Zealand, Sue Thompson displaced mottled triplefins, *Forsterygion malcomi,* 700 m (that is roughly the length of 7 football fields) along the rocky reefs on which these fish lived. After 4-6 days, 8 out of 10 displaced triplefins had returned home. This is not bad for a fish that is only 7-10 cm (3-4 inches) long and that spends all of its adult life on territories that are only 2 m^2. Similar results have been obtained with radiated shannies displaced over 270 m, flathead catfish over 1 km, cardinalfishes over 1-2 km, various sunfishes and bass over 3.5 km, and yellowtail rockfish over 22 km.

All of this brings us to the key question: How do fishes find their way back home? A first possibility is the existence of internal compasses. These can be based on sun position, polarised light fields, magnetic fields, or electrical fields. Olfaction may also play an important role in home-finding: fishes could be drawn by the smell of their home. Finally, visual landmarks can provide beacons for orientation. As this list suggests, fishes can resort to a number of different mechanisms to find their way around, and we should bear in mind that these mechanisms are not necessarily exclusive of one another.

Sun Compass

Most people are familiar with the notion that, with a watch and a view of the sun, it is possible to infer the position of any cardinal point. The sun is always over the east in the morning, over the south at midday (north in the Southern Hemisphere), and over the west at the end of the day. Points in between can be interpolated. Because fishes have an internal (circadian) clock that allows them to estimate the time of day, they can use the position of the sun to infer cardinal directions. Classical tests go like this: first, a fish is placed inside a container in the middle of a circular pool. The surroundings are uniform except for a view of the sun, or at the very least a bright lamp that moves around like the sun. The pool itself is as uniform as possible, and it is regularly rotated to prevent the fish from learning to rely on small landmarks inside the pool that might not be perceivable by people.

All around the periphery of the pool are a number of identical shelters or feeding stations, depending on the motivation of the fish at the moment of the test (finding shelter or finding food). They too can be rotated regularly. It is assumed that the fish always wants to go in the same cardinal direction, either because it does so in nature (some fish, for example, go offshore to feed and inshore to take shelter, and these represent constant directions throughout the life of a fish whose home range remains the same), or because the individual has been previously trained by the researcher to always go in the same direction (for example, of all the shelters around the periphery, only the one to the south-east is open).

So, the fish is released from the central container and the direction in which it swims is noted. If the fish can use sun-compass orientation, it should always swim in the correct direction, no matter what time of day it is, just as long as the sun (or lamp) has moved around the pool at a natural rate. Machiavellian researchers can also hide the true position of the sun and use mirrors to deflect its apparent position by, say, 90o, with the expectation that the preferred direction of the fish would also shift by 90o. Another variant is to test the fish with and without a view of the sun (on sunny and cloudy days, for example) with the expectation that correct orientation would be lost when the sun is not visible. With these methods, sun-compass orientation has been demonstrated in at least a dozen species of fish: white bass, pumpkinseed sunfish, bluegill sunfish, green sunfish, largemouth bass, Southern starhead topminnow, sockeye salmon, mosquitofish, two cichlids (*Cichlaurus (=Heros?) severus* and the uaru), one characid, and two parrotfishes (the purple and the rainbow parrotfish). In at least one of those studies, the contribution of the internal circadian clock was convincingly illustrated. Phillip Goodyear and David Bennett, then at the Savannah River Ecology Laboratory in South Carolina, captured immature bluegill sunfish that were in the habit of moving in a known direction to reach their natural refuge. When tested in a circular pool at midday in full view of the sun, these fish oriented correctly, equating the sun position with a southerly direction and moving appropriately relative to that. But Goodyear and Bennett also kept some fish in the lab under a photoperiod that had been advanced by 6 h (the fish "got up" at 1 AM instead of 7 AM).

When these fish were tested under the natural midday sun, they did not orient as if the sun was over the south. Instead they interpreted the sun position as being west. Their clock, which had been advanced along with the artificial photoperiod, told them they had been up for 11 hours and that this was the end of the day, and every decent fish knows that the sun is over the west at the end of the day! These results clearly indicated that fishes do not use sun height or even light colour as temporal cues, but sun position and the counsel of their internal clock instead. Sun-compass orientation has some limitations in aquatic environments. Objects that are low over the horizon cannot be seen from within the water, because of the reflection of light rays at the water surface. The sun, therefore, cannot be viewed by fish around dawn and dusk. Moreover, the sun disk can be hidden by clouds and, even if it is not, its precise position remains hard to establish from depths of more than a few metres. Some attention has therefore been devoted to indirect means of detecting the sun's position, means that overcome the above limitations.

If fact, there is only one good possibility for this, and it is the ability to perceive and orient to the pattern of light polarisation that is present in the sky. The alignment of this polarised light field varies predictably with the position of the sun, can be seen from depths of 200 m or more (albeit in reduced intensity), and can be seen in open parts of the sky even when the sun disk itself is low over the horizon or obscured by clouds. While he was at McMaster University in Hamilton, Ontario, Craig Hawryshyn tested the ability of rainbow trout to orient to polarised light. He and his co-workers first created a polarised light field by fitting linear polarising filters to a tungsten-halogen projector hanging from the ceiling. Then they placed a long rectangular aquarium parallel to the axis (the so-called "e-vector") of that light field, and trained trout to swim along this pathway to reach a shelter at the end. Then, in the same light field, they set up a circular wading pool, released the trained trout in the middle of it, and noted the direction in which the fish swam. The majority of the trout still moved parallel to the field vector, even though the circular pool now allowed them to go in any other direction.

The experiment worked again when other fish were trained to swim perpendicular to the vector rather than parallel to it. These fish maintained the learned perpendicular orientation when tested in the circular pool. Obviously, for them to do this, they had to be able to perceive the orientation of the polarised light field.

Such results indicate that fishes may be capable of using a polarised light compass (or, in other words, a sun compass with an indirect means of pinpointing sun position). However, the true usefulness of such a compass in nature is still uncertain. The Hawryshyn team has also shown that polarised light detection requires the presence of UV-sensitive cones in the retinas of the trout, and these cones may be lost when the fish become adult. Indeed, in the above experiments the researchers could not get adult trout to orient properly, only immatures of less than 30 g. The use of polarised light compasses may therefore be restricted only to those fish species that can perceive UV light. A further limitation resides in the progressive degradation of the polarised light field as depth and turbidity increase. In a subsequent study by Hawryshyn, less than 20% of tested trout oriented properly under a field that was only 65% polarised. Therefore, even though polarised light can be detected at great depth, its poor quality there may prevent its use as a spatial cue. The topic of orientation by polarised light needs to be further studied, preferably under the natural sky.

Magnetic and Electric Compasses

We are familiar with the magnetic compass, the kind we use when we go on backcountry hikes. The magnetized needle points north. A similar mechanism can be used by animals. Magnetite particles, Fe3O4, which can form the core of biological magnetic receptors, have been found in the head of salmon 30 and inside specialised cells within the nose of trout. These cells were found to be connected to the brain by a special nerve, and signals were detected on this nerve when the fish perceived magnetic anomalies in their environment. Fishes in fact represent the first vertebrate for which such potential magnetoreceptors *and* associated neurological hardware have been identified. This anatomical evidence is rather recent, and researchers did not wait for it to prove that fishes could maintain their bearings in a magnetic field. Protocols for this task were simple enough, although they necessitated the use of equipment capable of creating artificial magnetic fields or modifying the natural one.

Once so equipped, all that was left to do was to obtain fish that showed a consistent orientation in a magnetic field – either naturally or because they had been trained to do so – and then alter the field. One expectation was that the initial orientation would disappear if the magnetic field was abolished. And if the magnetic field was experimentally rotated, the prediction was that the preferred orientation of the fish would also rotate. In many cases, such predictions were upheld. Lincoln Chew and Grant Brown, at the University of Lethbridge, Alberta, put commercially-bred rainbow trout in circular arenas, and found that these fish repeatedly (but unexplainably) faced towards the magnetic north. The trout were not lining up with spatial cues from outside the pool because the visible surroundings were regularly moved between trials, and yet the fish still faced north. However, when the experiment was repeated in a room enclosed in Mu-metal (a special nickelsilver alloy) and lined with grounded copper strips to eliminate the natural magnetic field, the orientation of the trout became random. Similar results have been reported in leopard sharks, *Triakis semifasciata,* which also, intriguingly, oriented in a northerly direction when placed in a normal magnetic field. At the University of Auckland in New Zealand, P.B. Taylor kept introduced Chinook salmon in a rectangular tank oriented east-west, with the water flow and the food coming from the west. The fish, understandably, faced west. Things got more interesting when the salmon were moved, 18 months later, to a circular tank in another location without a view of the sky but within a normal geomagnetic field. Even though the water flow was not directional any more, the fish still faced west.

Taylor installed Helmholtz coils (circular wraps of copper wire energised by a DC power supply) around the arena and, by turning on the power in the appropriate coils, eliminated any trace of a magnetic field. The orientation of the fish became random. In other trials, the coils were realigned so as to preserve the local magnetic field, but rotated 90o clockwise. The fish's orientation also changed by 90o, but either clockwise or counter-clockwise (they faced north or south instead of west). Because the salmon in the shifted field faced either north or south rather than north only, their behaviour was called alignment rather than orientation. There is no solid explanation for this quirk of behaviour. Perhaps alignment results from an imperfect detection of the magnetic field polarity. At any rate, alignment seems to be peculiar to magnetic orientation. It is not seen in other types of compass.

This is illustrated by the work of Thomas Quinn and Ernest Brannon from the University of Seattle, who studied the migration behaviour of juvenile sockeye salmon in Babine Lake, British Columbia.

This lake is 150 km long and lies along a northwest-southeast axis. At the time of their capture, the young smolt studied by Quinn and Brannon were intent on leaving the lake on their seaward journey. When tested in circular arenas, they oriented properly in one direction: towards the outlet of the lake in the northwest. This, however, happened only when they had a clear view of the sky, and at such times their performance was not affected by the orientation of the magnetic field (normal or rotated 90o counterclockwise). However, when opaque covers were draped over the arenas to mask the view of the sky, the fish started to orient both towards and straight away from the outlet, and they readjusted this alignment accordingly when the magnetic field was rotated.

These findings first reveal a hierarchy of compasses, with the visual compass having priority over the magnetic one. When the sun is visible, only the sun compass is used, and proper orientation without alignment is achieved. When the position of the sun cannot be established, the fish is forced to rely on the magnetic compass, and alignment rather than complete orientation is the result. This is rather bad for those fish that end up facing away from the outlet of the lake, but perhaps they can eventually use other cues, such as water currents, to realise the error of their ways. Another, more esoteric kind of compass is an electrical one. Several fish species can detect electrical fields. In one experiment, a circular tank was set up with electrodes at 90o intervals along the periphery. When a pair of diametrically-opposed electrodes was activated, a DC current passed uniformly through the tank. In this arena, brown bullheads, *Ameiurus nebulosus*, were trained to swim from a central shelter to a feeding station 45o to the right of the negative electrode, a task they learned quickly. Now, when the electrodes were suddenly switched off and the other pair was turned on, the catfish searched for food 45o to the right of the newly-active negative electrode, not the previous one. Electrical compasses are therefore a reality, at least in catfish. Whether this means of orientation is useful to catfish in nature would depend on the existence of steady voltage gradients in their environment. Such hydroelectric fields have indeed been measured in some ponds and streams.

Use of Olfaction

All compasses, be they visual, magnetic, or electrical, provide only partial information. They inform fishes about the heading that should be followed in order to Bget home, but not the precise location of home.

If a sunfish, for example, learns through regular short-range excursions that its home is on the eastern shore of a lake, then after a long displacement it can use its compass and head east until it hits the shore. Once there however, it may still have to swim up and down the shoreline at random before it finally stumbles upon its familiar surroundings. This could take time, especially if the eastern shore is a long straight one. Similar scenarios can be imagined for coastal species. Are there any other tricks the fish can resort to so as to facilitate their task once their compass ceases to be useful? A good mechanism would be to cue in to the smell of the home area.

Such a smell would normally radiate in all directions if there is no current, or it would form a plume downstream if there are currents. Upon hitting a familiar smell, the fish could be induced to swim up the gradient of odour, or up current if there is one, until it finds the visual landmarks or the normal odour concentration of its home area. This idea assumes that the home area is endowed with a different odour than all other regions, that fish can learn this odour, and that their sense of smell is good enough to detect it even at very diffuse concentrations. All of this would be proven if anosmic (smellimpaired 40) fish could be shown to have reduced homing success as compared to normal individuals. Several studies have in fact managed to do this. On the west coast of Vancouver Island, Hong Woo Khoo, a PhD student in the laboratory of Norman Wilimovsky at the University of British Columbia, studied the homing capacity of the tidepool sculpin, *Oligocottus maculosus*, within its intertidal habitat. He captured many individuals from their home pools and made some of them anosmic by cauterising the inside of their nose while leaving the other fish untouched.

Then he displaced all of them up to 125 m along the shoreline. Less than 8% of the smell-impaired fish found their way back to the home pool, versus 20-80% – depending on the experiment – for the intact fish. The poor success of the anosmic fish was not caused by post-operative shock, because other sculpins were also subjected to surgery, this time to make them blind, and these blind fish homed almost as well as the controls. They could still smell, and that seemed to be the key .41 In a similar experiment, Morten Halvorsen and Ole Stabell from the University of Tromso in Norway displaced brown trout 200 m upstream or downstream from their home site. Beforehand, they had anaesthetised the fish and cauterised the olfactory detectors of some of them while cauterising two sites near the nasal openings of the others (this latter procedure did not make the fish anosmic but provided a control for the possible shock of operation and handling). More than four times as many control fish came back to their home within the next 9 weeks as compared to the smellimpaired individuals. Interestingly, the controls that had been displaced upstream, where they could not smell home, came back just as successfully as those that had been moved downstream.

We can therefore imagine that, within a stream habitat, displaced fish could follow a rule of thumb such as "If you can smell home, swim against the current, but if what you smell is not like home, then swim with the current until you do smell home". Most of the anosmic trout were recaptured around their site of release, so maybe their rule of thumb was "If you can't smell anything, stay put". It is a short jump from trout to salmon, which portray the classical example for the importance of olfaction in homing behaviour. When they are young, salmon learn the smell of the stream in which they live. Later, they leave the stream to go live at sea. Much later still, they come back to their natal stream in order to spawn in it. These adult salmon find their natal stream by following their nose – the trail of the odour learned several years earlier but not forgotten. The main player in the research that elucidated the mystery of this homing mechanism is Arthur Hasler from the University of Wisconsin. The time frame is the 1960s and 70s. Hasler's first experiment followed the now familiar protocol of catching adult salmon, making some of them anosmic (by plugging their nostrils) while leaving others untouched, and measuring their respective success at returning to the stream where they had been born and originally tagged. As you can guess, fish with unplugged noses made it home successfully whereas the anosmic ones were recaptured more or less evenly among all of the streams of the basin. The point was driven home (along with the fish) in a subsequent series of elegant experiments. Hasler and his co-workers reared young coho salmon in a hatchery and exposed them to one of two different chemicals, morpholine and phenethyl alcool (PEA). These artificial chemicals do not normally carry biological meaning but they are odoriferous. The fish were then marked according to the chemical they had been exposed to, and released into Lake Michigan. During the spawning migration 1.5 years later, the researchers dripped morpholine into one river and PEA into another 9 km away. Convincingly, 95% of the fish that were recaptured and that had been exposed to morpholine were recovered in the morpholine-scented river, and 92% of the recaptured PEA fish were recovered in the PEA-scented stream.

One cannot ask for a better experimental demonstration of the importance of odours for homing salmon, We still do not know the nature of the chemicals that provide the odour learned by fishes in the wild. Geosmin, a chemical produced by tiny mushrooms and present within inland waters, is a possibility. Glass eels have been shown to detect it and to prefer water laced with it at a time when they want to migrate into rivers. Another intriguing possibility is that the smell of other fish might contribute to the olfactor "bouquet" of the home stream. Experiments have revealed that salmon can distinguish between the smell of conspecifics from their own population versus that of others. One of these experiments showed that Arctic char which were reared in hatcheries with some of their brothers and sisters, the smell of which could therefore be memorised, preferred, once released in the wild, to ascend the river where other relatives were present. So maybe adult salmon migrating upriver could seek the smell of the juveniles presently living in their natal stream, assuming that these adults and juveniles all share an ancestral olfactory signature. Many coral reef fishes go through a drifting larval stage before settling down on a patch of coral.

A number of studies have shown that the presence of other residents on coral heads can influence the probability that larvae will settle there. In one field study conducted by Hugh Sweatman on the Great Barrier Reef of Australia, water was pumped at night from coral heads with and without resident humbug damselfish, into similar but unoccupied coral patches. The larval fish that settled through the night on these scented patches were collected at dawn by spreading the anaesthetic quinaldine over the coral. At the sites supplied with humbug-occupied coral water, many larvae of the humbug damsels had settled. At the sites supplied with unoccupied coral water, fewer larvae had settled. Sweatman concluded that the odour of adult conspecifics acts as an attractant in the settlement behaviour of larval humbugs. He pointed out that chemical information may be important for all species, such as humbug damselfish, that settle at night and cannot visually inspect their prospective real estate. Some experimenters have reported that larval coral fish, once released at night in open waters 1 km away from the nearest reef, unerringly take off in the direction of that reef. Perhaps they follow an odour plume emanating from the reef. A recent study has found that the larvae of two coral reef species – the spiny chromis *Acanthochromis polyacanthus*, and a cardinalfish – can distinguish between water from their own natal reef and water from a foreign reef, showing a preference for the natal reef water. Whether the relevant odour comes from the reef itself or from its occupants is unknown.

Use of Sounds

The idea has also been put forward that larval coral fish use sounds made by crashing waves and by the adult fish already on the reef as a directional cue. Some evidence for this was provided in 2005 when a paper in Science magazine reported that more fish larvae – mostly cardinalfishes and damselfishes – settled on reefs on which speakers broadcasting sounds of snapping shrimp and fish calls had been installed, as opposed to "silent" reefs.

Use of Landmarks

A final possibility for orientation is the recognition of landmarks. 52 While it seems likely that familiar landmarks are used by fishes for home range recognition, their usefulness for long-distance orientation is more doubtful given the limited visibility that can be achieved in water. But for some type of short-distance movement, landmarks could be helpful. The behaviour of the small frillfin goby *Bathygobius soporator* is interesting in that respect. These fish live in the intertidal zone and at low tide they are confined to tide pools. If they are chased by mad scientists while in such a pool, they jump out of the pool and "land", with amazing accuracy, in adjacent pools! Sometimes they jump from pool to pool until they reach open water, a trip that may require up to 6 different jumps, not all of them in the same direction.

This works only when the fish have had a chance to explore the whole area at high tide, when all pools are covered by water and swimming between them is possible. When introduced into an unfamiliar pool at low tide, gobies refuse to jump or they jump wrongly onto rocks. But after only one night of exploring around that new pool at high tide, the jumping behaviour becomes accurate again. This is true even under cloudy skies, which shows that a sun compass is not used. There is still the possibility of a magnetic compass at work, but the most likely hypothesis is that the shape of each pool is memorised and serves as the main cue for proper orientation towards the next landing place. Fishes could also learn the order of a series of landmarks and establish routes along them. Many coral reef fishes move off the reef on their way to feeding areas at dawn and back along the same route at dusk.

Every day, the same route is used. Divers who have observed this behaviour could not help but reflect that the fish were following a series of specific landmarks. Supporting this idea, individual butterflyfish that were experimentally displaced off the route seemed to move at random at first, but as soon as they happened to cross their well-known path, they turned and went straight home along it, as if recognising it. Unfortunately, observations of this kind do not provide information about the exact nature of the specific landmarks. To learn more, one would have to remove some of the possible landmarks. Such a task would be made difficult by the fact that relatively large areas of sand or rock might be used as signposts. These would be hard to remove! Whatever the mechanism or species involved, successful homing by animals inspires awe in anyone who witnesses it – and witness it we can if we take the time to sit by a tide pool and recognise the same sculpin in it, or by a turbulent stream and admire the efforts of a salmon ascending it. There is more than a little mystery about the art of not getting lost, but a good part of it has been made less puzzling by patient research.

This field of investigation is difficult, and therefore stimulating for some, because it requires broad knowledge in sensory physiology (vision, olfaction, electroreception, magnetoreception), chronobiology (circadian clocks), and physics (magnetic fields, celestial movement), not to mention expertise in the capture and tagging of wild animals. But even knowing how fish succeed in homing does not make the feat less awesome, as the mechanisms themselves turn out to be exquisitely complex. They are also remarkably efficient in the face of the aquatic environment's vast expanses.

Chapter 4

Fish Senses

Fish are a paraphyletic group of organisms that consist of all gill-bearing aquatic vertebrate (or craniate) animals that lack limbs with digits. Included in this definition are the living hagfish, lampreys, and cartilaginous and bony fish, as well as various extinct related groups. Most fish are ectothermic ("cold-blooded"), allowing their body temperatures to vary as ambient temperatures change, though some of the large active swimmers like white shark and tuna can hold a higher core temperature. Fish are abundant in most bodies of water. They can be found in nearly all aquatic environments, from high mountain streams (e.g., char and gudgeon) to the abyssal and even hadal depths of the deepestoceans (e.g., gulpers and anglerfish). At 32,000 species, fish exhibit greater species diversity than any other class of vertebrates. Fish, especially as food, are an important resource worldwide.

Commercial and subsistence fishers hunt fish in wild fisheries orfarm them in ponds or in cages in the ocean. They are also caught by recreational fishers, kept as pets, raised byfishkeepers, and exhibited in public aquaria. Fish have had a role in culture through the ages, serving as deities, religious symbols, and as the subjects of art, books and movies. Because the term "fish" is defined negatively, and excludes the tetrapods (i.e., the amphibians, reptiles, birds and mammals) which descend from within the same ancestry, it is paraphyletic, and is not considered a proper grouping in systematic biology. The traditional term pisces(also ichthyes) is considered a typological, but not a phylogenetic classification.

The earliest organisms that can be classified as fish were soft-bodied chordates that first appeared during the Cambrian period. Although they lacked a true spine, they possessed notochords which allowed them to be more agile than their invertebrate counterparts. Fish would continue to evolve through the Paleozoic era, diversifying into a wide variety of forms. Many fish of the Paleozoic developed external armour that protected them from predators. The first fish with jaws appeared in the Silurian period, after which many (such as sharks) became formidable marine predators rather than just the prey of arthropods.

Diversity of Fish

Figure: Fish come in many shapes and sizes. This is a sea dragon, a close relative of theseahorse. Their leaf-like appendages enable them to blend in with floating seaweed.

The term "fish" most precisely describes any non-tetrapod craniate (i.e. an animal with a skull and in most cases a backbone) that has gillsthroughout life and whose limbs, if any, are in the shape of fins. Unlike groupings such as birds or mammals, fish are not a single clade but aparaphyletic collection of taxa, including hagfishes, lampreys, sharks and rays, ray-finned fish, coelacanths, and lungfish. Indeed, lungfish and coelacanths are closer relatives of tetrapods (such as mammals, birds, amphibians, etc.) than of other fish such as ray-finned fish or sharks, so the last common ancestor of all fish is also an ancestor to tetrapods. As paraphyletic groups are no longer recognised in modernsystematic biology, the use of the term "fish" as a biological group must be avoided.

Many types of aquatic animals commonly referred to as "fish" are not fish in the sense given above; examples include shellfish, cuttlefish, starfish, crayfish and jellyfish. In earlier times, even biologists did not make a distinction – sixteenth century natural historians classified alsoseals, whales, amphibians, crocodiles, even hippopotamuses, as well as a host of aquatic invertebrates, as fish. However, according the definition above, all mammals, including cetaceans like whales and dolphins, are not fish. In some contexts, especially in aquaculture, the true fish are referred to as finfish (or fin fish) to distinguish them from these other animals. A typical fish is ectothermic, has a streamlined body for rapid swimming, extracts oxygen from water using gills or uses an accessory breathing organ to breathe atmospheric oxygen, has two sets of paired fins, usually one or two (rarely three) dorsal fins, an anal fin, and a tail fin, has jaws, has skin that is usually covered with scales, and lays eggs.

Each criterion has exceptions. Tuna, swordfish, and some species of sharks show some warm-blooded adaptations—they can heat their bodies significantly above ambient water temperature. Streamlining and swimming performance varies from fish such as tuna, salmon, and jacks that can cover 10–20 body-lengths per second to species such as eels andrays that swim no more than 0.5 body-lengths per second. Many groups of freshwater fish extract oxygen from the air as well as from the water using a variety of different structures.Lungfish have paired lungs similar to those of tetrapods, gouramis have a structure called the labyrinth organ that performs a similar function, while many catfish, such as *Corydoras*extract oxygen via the intestine or stomach. Body shape and the arrangement of the fins is highly variable, covering such seemingly un-fishlike forms as seahorses, pufferfish,anglerfish, and gulpers. Similarly, the surface of the skin may be naked (as in moray eels), or covered with scales of a variety of different types usually defined as placoid (typical of sharks and rays), cosmoid (fossil lungfish and coelacanths), ganoid (various fossil fish but also living gars and bichirs), cycloid, and ctenoid (these last two are found on most bony fish). There are even fish that live mostly on land. Mudskippers feed and interact with one another on mudflats and go underwater to hide in their burrows. The catfish *Phreatobius cisternarum* lives in underground, phreatic habitats, and a relative lives in waterlogged leaf litter.

Fish range in size from the huge 16-metre (52 ft) whale shark to the tiny 8-millimetre (0.3 in) stout infantfish.

Fish species diversity is roughly divided equally between marine (oceanic) and freshwater ecosystems. Coral reefs in the Indo-Pacific constitute the centre of diversity for marine fishes, whereas continental freshwater fishes are most diverse in large river basins of tropical rainforests, especially the Amazon, Congo, and Mekong basins. More than 5,600 fish species inhabit Neotropical freshwaters alone, such that Neotropical fishes represent about 10% of all vertebrate species on the Earth.

Taxonomy

Fish are a paraphyletic group: that is, any clade containing all fish also contains the tetrapods, which are not fish. For this reason, groups such as the "Class Pisces" seen in older reference works are no longer used in formal classifications. Traditional classification divide fish into three extant classes, and with extinct forms sometimes classified within the tree, sometimes as their own classes:

- Class Agnatha (Jawless fish)
  - o Subclass Cyclostomata (hagfish and lampreys)
  - o Subclass Ostracodermi (armoured jawless fish) †
- Class Chondrichthyes (cartilaginous fish)
  - o Subclass Elasmobranchii (sharks and rayes)
  - o Subclass Holocephali (chimaeras and extinct relatives)
- Class Placodermi (armoured fish) †
- Class Acanthodii ("spiny sharks", sometimes classified under bony fishes)†
- Class Osteichthyes (bony fish)
  - o Subclass Actinopterygii (ray finned fishes)
  - o Subclass Sarcopterygii (fleshy finned fishes, ancestors of tetrapods).

The above scheme is the one most commonly encountered in non-specialist and general works. Many of the above groups are paraphyletic, in that they have given rise to successive groups: Agnathans are ancestral to Chondrichthyes, who again have given rise to Acanthodiians, the ancestors of Osteichthyes. With the arrival of phylogenetic nomenclature, the fishes has been split up into a more detailed scheme, with the following major groups:

- Class Myxini (hagfish)
- Class Pteraspidomorphi † (early jawless fish)
- Class Thelodonti
- Class Anaspida
- Class Petromyzontida or Hyperoartia

- Petromyzontidae (lampreys)
- Class Conodonta (conodonts) †
- Class Cephalaspidomorphi † (early jawless fish)
    - (unranked) Galeaspida †
    - (unranked) Pituriaspida †
    - (unranked) Osteostraci †
- Infraphylum Gnathostomata (jawed vertebrates)
    - Class Placodermi † (armoured fish)
    - Class Chondrichthyes (cartilaginous fish)
    - Class Acanthodii † (spiny sharks)
    - Superclass Osteichthyes (bony fish)
- Class Actinopterygii (ray-finned fish)
- Subclass Chondrostei
- Order Acipenseriformes (sturgeons and paddlefishes)
- Order Polypteriformes (reedfishes and bichirs).
- Subclass Neopterygii
- Infraclass Holostei (gars and bowfins)
- Infraclass Teleostei (many orders of common fish)
- Class Sarcopterygii (lobe-finned fish)
- Subclass Coelacanthimorpha (coelacanths)
- Subclass Dipnoi (lungfish).

Some palaeontologists contend that because Conodonta are chordates, they are primitive fish. The position of hagfish in the phylum chordata is not settled. Phylogenetic research in 1998 and 1999 supported the idea that the hagfish and the lampreys form a natural group, theCyclostomata, that is a sister group of the Gnathostomata.

The various fish groups account for more than half of vertebrate species. There are almost 28,000 known extant species, of which almost 27,000 are bony fish, with 970 sharks, rays, and chimeras and about 108 hagfish and lampreys.

A third of these species fall within the nine largest families; from largest to smallest, these families are Cyprinidae, Gobiidae, Cichlidae, Characidae, Loricariidae, Balitoridae, Serranidae, Labridae, and Scorpaenidae. About 64 families are monotypic, containing only one species. The final total of extant species may grow to exceed 32,500.

Anatomy

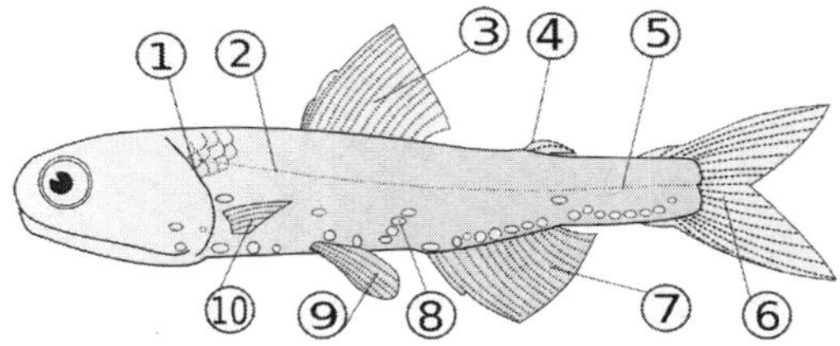

Figure: The anatomy of *Lampanyctodes hectoris;* (1) operculum (gill cover), (2) lateral line, (3) dorsal fin, (4) fat fin, (5) caudal peduncle, (6) caudal fin, (7) anal fin, (8) photophores, (9) pelvic fins (paired), (10) pectoral fins (paired)

Respiration

Most fish exchange gases using gills on either side of the pharynx. Gills consist of threadlike structures calledfilaments. Each filament contains a capillary network that provides a large surface area for exchanging oxygen andcarbon dioxide. Fish exchange gases by pulling oxygen-rich water through their mouths and pumping it over their gills. In some fish, capillary blood flows in the opposite direction to the water, causing countercurrent exchange. The gills push the oxygen-poor water out through openings in the sides of the pharynx. Some fish, like sharks and lampreys, possess multiple gill openings. However, bony fish have a single gill opening on each side. This opening is hidden beneath a protective bony cover called an operculum.

Juvenile bichirs have external gills, a very primitive feature that they share with larval amphibians.

Fish from multiple groups can live out of the water for extended time periods. Amphibious fish such as the mudskippercan live and move about on land for up to several days, or live in stagnant or otherwise oxygen depleted water. Many such fish can breathe air via a variety of mechanisms. The skin of anguillid eels may absorb oxygen directly. The buccal cavity of the electric eel may breathe air. Catfish of the families Loricariidae, Callichthyidae, and Scoloplacidae absorb air through their digestive tracts. Lungfish, with the exception of the Australian lungfish, and bichirs have paired lungs similar to those of tetrapods and must surface to gulp fresh air through the mouth and pass spent air out through the gills. Gar andbowfin have a vascularised swim bladder that functions in the same way. Loaches, trahiras, and many catfish breathe by passing air through the gut. Mudskippers breathe by absorbing oxygen across the skin (similar to frogs). A number of fish have evolved so-called accessory breathing organs that extract oxygen from the air. Labyrinth fish (such as gouramis andbettas) have a labyrinth organ above the gills that performs this function. A few other fish have structures resembling labyrinth organs in form and function, most notably snakeheads,pikeheads, and the Clariidae catfish family.

Breathing air is primarily of use to fish that inhabit shallow, seasonally variable waters where the water's oxygen concentration may seasonally decline. Fish dependent solely on dissolved oxygen, such as perch and cichlids, quickly suffocate, while air-breathers survive for much longer, in some cases in water that is little more than wet mud. At the most extreme, some air-breathing fish are able to survive in damp burrows for weeks without water, entering a state of aestivation (summertime hibernation) until water returns.

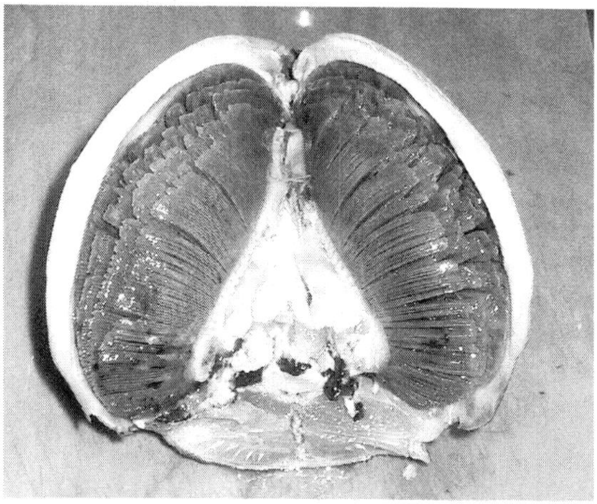

Figure: Tuna gills inside of the head. The fish head is oriented snout-downwards, with the view looking towards the mouth.

Air breathing fish can be divided into obligate air breathers and facultative air breathers. Obligate air breathers, such as theAfrican lungfish, *must* breathe air periodically or they suffocate. Facultative air breathers, such as the catfish *Hypostomus plecostomus*, only breathe air if they need to and will otherwise rely on their gills for oxygen. Most air breathing fish are facultative air breathers that avoid the energetic cost of rising to the surface and the fitness cost of exposure to surface predators.

Circulation

Fish have a closed-loop circulatory system. The heart pumps the blood in a single loop throughout the body. In most fish, the heart consists of four parts, including two chambers and an entrance and exit. The first part is the sinus venosus, a thin-walled sac that collects blood from the fish's veins before allowing it to flow to the second part, the atrium, which is a large muscular chamber. The atrium serves as a one-way antechamber, sends blood to the third part, ventricle. The ventricle is another thick-walled, muscular chamber and it pumps the blood, first to the fourth part, bulbus arteriosus, a large tube, and then out of the heart. The bulbus arteriosus connects to the aorta, through which blood flows to the gills for oxygenation.

Digestion

Jaws allow fish to eat a wide variety of food, including plants and other organisms. Fish ingest food through the mouth and break it down in the esophagus. In the stomach, food is further digested and, in many fish, processed in finger-shaped pouches calledpyloric caeca, which secrete digestive enzymes and absorb nutrients. Organs such as the liver and pancreas add enzymes and various chemicals as the food moves through the digestive tract. The intestine completes the process of digestion and nutrient absorption.

Excretion

As with many aquatic animals, most fish release their nitrogenous wastes as ammonia. Some of the wastes diffuse through the gills. Blood wastes are filtered by the kidneys.

Saltwater fish tend to lose water because of osmosis. Their kidneys return water to the body. The reverse happens in freshwater fish: they tend to gain water osmotically. Their kidneys produce dilute urine for excretion. Some fish have specially adapted kidneys that vary in function, allowing them to move from freshwater to saltwater.

Scales

The scales of fish originate from the mesoderm (skin); they may be similar in structure to teeth.

Sensory and Nervous System

Central Nervous System: Fish typically have quite small brains relative to body size compared with other vertebrates, typically one-fifteenth the brain mass of a similarly sized bird or mammal. However, some fish have relatively large brains, most notably mormyrids and sharks, which have brains about as massive relative to body weight as birds and marsupials.

Fish brains are divided into several regions. At the front are the olfactory lobes, a pair of structures that receive and process signals from the nostrils via the two olfactory nerves. The olfactory lobes are very large in fish that hunt primarily by smell, such as hagfish, sharks, and catfish. Behind the olfactory lobes is the two-lobed telencephalon, the structural equivalent to the cerebrum in higher vertebrates. In fish the telencephalon is concerned mostly with olfaction. Together these structures form the forebrain.

Connecting the forebrain to the midbrain is the diencephalon (in the diagram, this structure is below the optic lobes and consequently not visible). The diencephalon performs functions associated with hormones and homeostasis. The pineal body lies just above the diencephalon. This structure detects light, maintains circadian rhythms, and controls colour changes. The midbrain or mesencephalon contains the two optic lobes. These are very large in species that hunt by sight, such as rainbow trout and cichlids.

The hindbrain or metencephalon is particularly involved in swimming and balance. The cerebellum is a single-lobed structure that is typically the biggest part of the brain. Hagfish and lampreys have relatively small cerebellae, while the mormyrid cerebellum is massive and apparently involved in their electrical sense. The brain stem or myelencephalon is the brain's posterior. As well as controlling some muscles and body organs, in bony fish at least, the brain stem governs respiration and osmoregulation.

Sense Organs

Most fish possess highly developed sense organs. Nearly all daylight fish have colour vision that is at least as good as a human's. Many fish also have chemoreceptors that are responsible for extraordinary senses of taste and smell. Although they have ears, many fish may not hear very well. Most fish have sensitive receptors that form the lateral line system, which detects gentle currents and vibrations, and senses the motion of nearby fish and prey. Some fish, such as catfish and sharks, have organs that detect weak electric currents on the order of millivolt. Other fish, like the South American electric fishes Gymnotiformes, can produce weak electric currents, which they use in navigation and social communication.

Fish orient themselves using landmarks and may use mental maps based on multiple landmarks or symbols. Fish behaviour in mazes reveals that they possess spatial memory and visual discrimination.

Vision

Vision is an important sensory system for most species of fish. Fish eyes are similar to those of terrestrial vertebrates like birds and mammals, but have a more spherical lens. Theirretinas generally have both rod cells and cone cells (for scotopic and photopic vision), and most species have colour vision. Some fish can see ultraviolet and some can see polarised light. Amongst jawless fish, the lamprey has well-developed eyes, while the hagfish has only primitive eyespots. Fish vision shows adaptation to their visual environment, for example deep sea fishes have eyes suited to the dark environment.

Capacity for Pain

Experiments done by William Tavolga provide evidence that fish have pain and fear responses. For instance, in Tavolga's experiments, toadfish grunted when electrically shocked and over time they came to grunt at the mere sight of an electrode.

In 2003, Scottish scientists at the University of Edinburgh and the Roslin Institute concluded that rainbow trout exhibit behaviours often associated with pain in other animals. Bee venomand acetic acid injected into the lips resulted in fish rocking their bodies and rubbing their lips along the sides and floors of their tanks, which the researchers concluded were attempts to relieve pain, similar to what mammals would do. Neurons fired in a pattern resembling human neuronal patterns.

Professor James D. Rose of the University of Wyoming claimed the study was flawed since it did not provide proof that fish possess "conscious awareness, particularly a kind of awareness that is meaningfully like ours". Rose argues that since fish brains are so different from human brains, fish are probably not conscious in the manner humans are, so that reactions similar to human reactions to pain instead have other causes. Rose had published a study a year earlier arguing that fish cannot feel pain because their brains lack aneocortex. However, animal behaviourist Temple Grandin argues that fish could still have consciousness without a neocortex because "different species can use different brain structures and systems to handle the same functions."

Animal welfare advocates raise concerns about the possible suffering of fish caused by angling. Some countries, such as Germany have banned specific types of fishing, and the British RSPCA now formally prosecutes individuals who are cruel to fish.

Muscular System

Most fish move by alternately contracting paired sets of muscles on either side of the backbone. These contractions form S-shaped curves that move down the body. As each curve reaches the back fin, backward force is applied to the water, and in conjunction with the fins, moves the fish forward. The fish's fins function like an airplane's flaps. Fins also increase the tail's surface area, increasing speed. The streamlined body of the fish decreases the amount of friction from the water. Since body tissue is denser than water, fish must compensate for the difference or they will sink. Many bony fish have an internal organ called a swim bladder that adjusts their buoyancy through manipulation of gases.

Figure: A 3-tonne (3.0-long-ton; 3.3-short-ton)great white shark off Isla Guadalupe

Homeothermy

Although most fish are exclusively ectothermic, there are exceptions. Certain species of fish maintain elevated body temperatures. Endothermic teleosts (bony fish) are all in the suborder Scombroidei and include the billfishes, tunas, and one species of "primitive" mackerel (*Gasterochisma melampus*). All sharks in the family Lamnidae – shortfin mako, long fin mako, white, porbeagle, and salmon shark – are endothermic, and evidence suggests the trait exists in family Alopiidae (thresher sharks). The degree of endothermy varies from the billfish, which warm only their eyes and brain, to bluefin tuna and porbeagle sharks who maintain body temperatures elevated in excess of 20 °C above ambient water temperatures. Endothermy, though metabolically costly, is thought to provide advantages such as increased muscle strength, higher rates of central nervous system processing, and higher rates of digestion.

Reproductive System

Organs

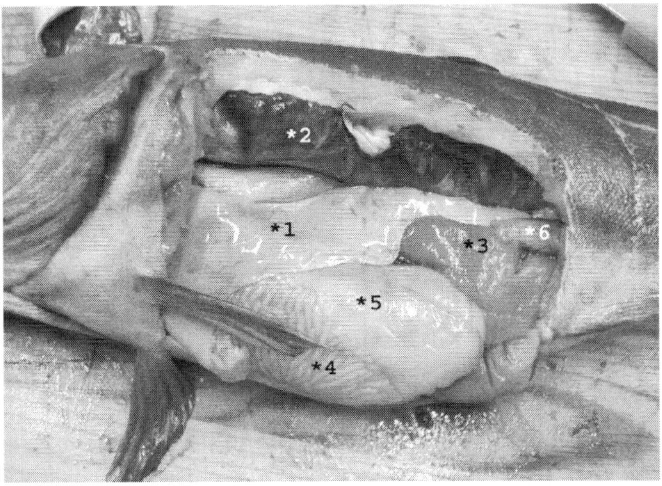

Figure: Organs: 1. Liver, 2. Gas bladder, 3. Roe, 4. Pyloric caeca, 5. Stomach, 6. Intestine

Fish reproductive organs include testes and ovaries. In most species, gonads are paired organs of similar size, which can be partially or totally fused. There may also be a range of secondary organs that increase reproductive fitness. In terms of spermatogonia distribution, the structure of teleosts testes has two types: in the most common, spermatogonia occur all along theseminiferous tubules, while in Atherinomorph fish they are confined to the distal portion of these structures. Fish can present cystic or semi-cystic spermatogenesis in relation to the release phase of germ cells in cysts to the seminiferous tubules lumen.

Fish ovaries may be of three types: gymnovarian, secondary gymnovarian or cystovarian. In the first type, the oocytes are released directly into the coelomic cavity and then enter the ostium, then through the oviduct and are eliminated. Secondary gymnovarian ovaries shed ova into thecoelom from which they go directly into the oviduct. In the third type, the oocytes are conveyed to the exterior through the oviduct.Gymnovaries are the primitive condition found in lungfish, sturgeon, and bowfin. Cystovaries characterise most teleosts, where the ovary lumen has continuity with the oviduct. Secondary gymnovaries are found in salmonids and a few other teleosts. Oogonia development in teleosts fish varies according to the group, and the determination of oogenesis dynamics allows the understanding of maturation and fertilization processes. Changes in the nucleus, ooplasm, and the surrounding layers characterise the oocyte maturation process.

Postovulatory follicles are structures formed after oocyte release; they do not have endocrine function, present a wide irregular lumen, and are rapidly reabsorbed in a process involving the apoptosis of follicular cells. A degenerative process called follicular atresia reabsorbs vitellogenic oocytes not spawned. This process can also occur, but less frequently, in oocytes in other development stages. Some fish are hermaphrodites, having both testes and ovaries either at different phases in their life cycle or, as in hamlets, have them simultaneously.

Reproductive Method

Over 97% of all known fish are oviparous, that is, the eggs develop outside the mother's body. Examples of oviparous fish include salmon, goldfish, cichlids, tuna, and eels. In the majority of these species, fertilisation takes place outside the mother's body, with the male and female fish shedding their gametes into the surrounding water. However, a few oviparous fish practice internal fertilization, with the male using some sort of intromittent organ to deliver sperm into the genital opening of the female, most notably the oviparous sharks, such as the horn shark, and oviparous rays, such as skates. In these cases, the male is equipped with a pair of modified pelvic fins known as claspers. Marine fish can produce high numbers of eggs which are often released into the open water column. The eggs have an average diametre of 1 millimetre (0.039 in).

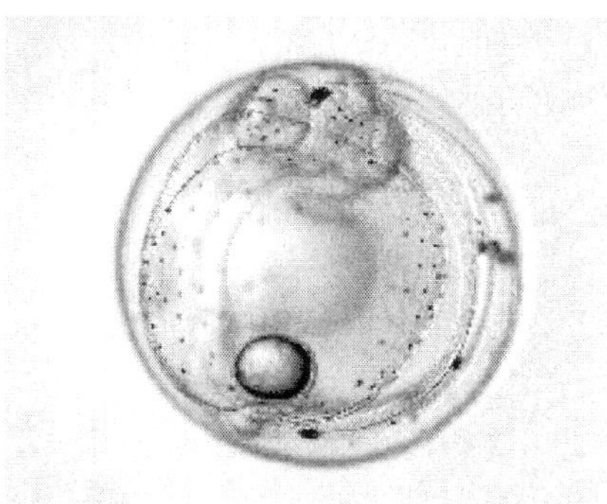

Figure: An example of zooplankton

The newly hatched young of oviparous fish are called larvae. They are usually poorly formed, carry a large yolk sac (for nourishment) and are very different in appearance from juvenile and adult specimens. The larval period in oviparous fish is relatively short (usually only several weeks), and larvae rapidly grow and change appearance and structure (a process termed metamorphosis) to become juveniles. During this transition larvae must switch from their yolk sac to feeding on zooplankton prey, a process which depends on typically inadequate zooplankton density, starving many larvae.

In ovoviviparous fish the eggs develop inside the mother's body after internal fertilization but receive little or no nourishment directly from the mother, depending instead on the yolk. Each embryo develops in its own egg. Familiar examples of ovoviviparous fish include guppies, angel sharks, and coelacanths.

Some species of fish are viviparous. In such species the mother retains the eggs and nourishes the embryos. Typically, viviparous fish have a structure analogous to the placenta seen in mammals connecting the mother's blood supply with that of the embryo. Examples of viviparous fish include the surf-perches, splitfins, and lemon shark. Some viviparous fish exhibit oophagy, in which the developing embryos eat other eggs produced by the mother. This has been observed primarily among sharks, such as the shortfin mako and porbeagle, but is known for a few bony fish as well, such as the halfbeak *Nomorhamphus ebrardtii*. Intrauterine cannibalism is an even more unusual mode of vivipary, in which the largest embryos eat weaker and smaller siblings. This behaviour is also most commonly found among sharks, such as the grey nurse shark, but has also been reported for *Nomorhamphus ebrardtii*. Aquarists commonly refer to ovoviviparous and viviparous fish as livebearers.

Immune System

Immune organs vary by type of fish. In the jawless fish (lampreys and hagfish), true lymphoid organs are absent. These fish rely on regions of lymphoid tissue within other organs to produce immune cells. For example, erythrocytes, macrophages and plasma cells are produced in the anterior kidney (or pronephros) and some areas of the gut (where granulocytesmature.) They resemble primitive bone marrow in hagfish. Cartilaginous fish (sharks and rays) have a more advanced immune system. They have three specialised organs that are unique to chondrichthyes; the epigonal organs (lymphoid tissue similar to mammalian bone) that surround the gonads, the Leydig's organ within the walls of their esophagus, and a spiral valve in their intestine. These organs house typical immune cells (granulocytes, lymphocytes and plasma cells). They also possess an identifiable thymus and a well-developed spleen (their most important immune organ) where various lymphocytes, plasma cells and macrophages develop and are stored. Chondrostean fish (sturgeons, paddlefish and bichirs) possess a major site for the production of granulocytes within a mass that is associated with the meninges (membranes surrounding the central nervous system.) Their heart is frequently covered with tissue that contains lymphocytes, reticular cells and a small number of macrophages. The chondrostean kidney is an important hemopoietic organ; where erythrocytes, granulocytes, lymphocytes and macrophages develop.

Like chondrostean fish, the major immune tissues of bony fish (or teleostei) include the kidney (especially the anterior kidney), which houses many different immune cells. In addition, teleost fish possess a thymus, spleen and scattered immune areas within mucosal tissues (e.g. in the skin, gills, gut and gonads). Much like the mammalian immune system, teleost erythrocytes, neutrophils and granulocytes are believed to reside in the spleen whereas lymphocytes are the major cell type found in the thymus. In 2006, a lymphatic system similar to that in mammals was described in one species of teleost fish, the zebrafish. Although not confirmed as yet, this system presumably will be where naive (unstimulated) T cellsaccumulate while waiting to encounter an antigen.

Diseases

Like other animals, fish suffer from diseases and parasites. To prevent disease they have a variety of defences. *Non-specific* defences include the skin and scales, as well as the mucus layer secreted by the epidermis that traps and inhibits the growth of microorganisms. If pathogens breach these defences, fish can develop an inflammatory response that increases blood flow to the infected region and delivers white blood cells that attempt to destroy pathogens. Specific defences respond to particular pathogens recognised by the fish's body, i.e., animmune response. In recent years, vaccines have become widely used in aquaculture and also with ornamental fish, for example furunculosis vaccines in farmed salmon and koi herpes virus in koi.

Some species use cleaner fish to remove external parasites. The best known of these are the Bluestreak cleaner wrasses of the genus *Labroides* found on coral reefs in the Indian andPacific Oceans. These small fish maintain so-called "cleaning stations" where other fish congregate and perform specific movements to attract the attention of the cleaners. Cleaning behaviours have been observed in a number of fish groups, including an interesting case between two cichlids of the same genus, *Etroplus maculatus*, the cleaner, and the much larger*Etroplus suratensis*.

Evolution

Proliferation of fish was apparently due to the hinged jaw, because jawless fish left very few descendants. Lampreys may approximate pre-jawed fish. The first jaws are found in Placodermi fossils. It is unclear if the advantage of a hinged jaw is greater biting force, improved respiration, or a combination of factors. Fish may have evolved from a creature similar to a coral-like Sea squirt, whose larvae resemble primitive fish in important ways. The first ancestors of fish may have kept the larval form into adulthood (as some sea squirts do today), although perhaps the reverse is the case.

Conservation

The 2006 IUCN Red List names 1,173 fish species that are threatened with extinction. Included are species such as Atlantic cod, Devil's Hole pupfish, coelacanths, and great white sharks. Because fish live underwater they are more difficult to study than terrestrial animals and plants, and information about fish populations is often lacking. However, freshwater fish seem particularly threatened because they often live in relatively small water bodies. For example, the Devil's Hole pupfish occupies only a single 3 by 6 metres (10 by 20 ft) pool.

Overfishing

Figure: A Whale shark, the world's largest fish, is classified as Vulnerable.

Overfishing is a major threat to edible fish such as cod and tuna. Overfishing eventually causes population (known as stock) collapse because the survivors cannot produce enough young to replace those removed. Such commercial extinction does not mean that the species is extinct, merely that it can no longer sustain a fishery.

One well-studied example of fishery collapse is the Pacific sardine *Sadinops sagax caerulues* fishery off the California coast. From a 1937 peak of 790,000 long tons (800,000 t) the catch steadily declined to only 24,000 long tons (24,000 t) in 1968, after which the fishery was no longer economically viable.

The main tension between fisheries science and the fishing industry is that the two groups have different views on the resiliency of fisheries to intensive fishing. In places such as Scotland, Newfoundland, and Alaska the fishing industry is a major employer, so governments are predisposed to support it. On the other hand, scientists and conservationists push for stringent protection, warning that many stocks could be wiped out within fifty years.

Habitat Destruction

A key stress on both freshwater and marine ecosystems is habitat degradation including water pollution, the building of dams, removal of water for use by humans, and the introduction of exotic species. An example of a fish that has become endangered because of habitat change is the pallid sturgeon, a North American freshwater fish that lives in rivers damaged by human activity.

Exotic Species

Introduction of non-native species has occurred in many habitats. One of the best studied examples is the introduction of Nile perch into Lake Victoria in the 1960s. Nile perch gradually exterminated the lake's 500 endemic cichlid species. Some of them survive now in captive breeding programmes, but others are probably extinct. Carp, snakeheads, tilapia,European perch, brown trout, rainbow trout, and sea lampreys are other examples of fish that have caused problems by being introduced into alien environments.

Importance to Humans

Culture

In the Book of Jonah a "great fish" swallowed Jonah the Prophet. Legends of half-human, half-fish mermaids have featured in stories like those of Hans Christian Andersen and movies like *Splash*. Among the deities said to take the form of a fish are Ika-Roa of the Polynesians, Dagon of various ancient Semitic peoples, the shark-gods of Hawaii and Matsya of the Hindus. The astrological symbol Pisces is based on a constellation of the same name, but there is also a second fish constellation in the night sky, Piscis Austrinus. Fish have been used figuratively in many different ways, for example the ichthys used by early Christians to identify themselves, through to the fish as a symbol of fertility among Bengalis. Fish feature prominently in art and literature, in movies such as *Finding Nemo* and books such as *The Old Man and the Sea*. Large fish, particularly sharks, have frequently been the subject of horror movies and thrillers, most notably the novel *Jaws*, which spawned a series of films of the same namethat in turn inspired similar films or parodies such as *Shark Tale*, *Snakehead Terror*, and *Piranha*.

In the semiotic of Ashtamangala (buddhist symbolism) the golden fish (Sanskrit: Matsya), represents the state of fearless suspension in samsara, perceived as the harmless ocean, referred to as 'buddha-eyes' or 'rigpa-sight'. The fish symbolises the auspiciousness of all living beings in a state of fearlessness without danger of drowning in the Samsaric Ocean of Suffering, and migrating from teaching to teaching freely and spontaneously just as fish swim. They have religious significance in Hindu, Jain and Buddhist traditions but also in Christianity who is first signified by the sign of the fish, and especially referring to feeding the multitude in the desert. In the dhamma of Buddha the fish symbolise happiness as they have complete freedom of movement in the water. They represent fertility and abundance. Often drawn in the form of carp which are regarded in the Orient as sacred on account of their elegant beauty, size and life-span. The name of the Canadian city of Coquitlam, British Columbia is derived from *Kwikwetlem*, which is said to be derived from a Coast Salish term meaning "little red fish".

Terminology

Shoal or School?

A random assemblage of fish merely using some localised resource such as food or nesting sites is known simply as an *aggregation*. When fish come together in an interactive, social grouping, then they may be forming either a *shoal* or a *school* depending on the degree of organisation. A *shoal* is a loosely organised group where each fish swims and forages independently but is attracted to other members of the group and adjusts its behaviour, such as swimming speed, so that it remains close to the other members of the group. *Schools* of fish are much more tightly organised, synchronising their swimming so that all fish move at the same speed and in the same direction.

Shoaling and schooling behaviour is believed to provide a variety of advantages.

Examples:

- Cichlids congregating at lekking sites form an *aggregation.*
- Many minnows and characins form *shoals.*
- Anchovies, herrings and silversides are classic examples of *schooling* fish.

While school and shoal have different meanings within biology, they are often treated as synonyms by non-specialists, with speakers of British English using "shoal" to describe any grouping of fish, while speakers of American English often using "school" just as loosely.

Fish or Fishes?

Though often used interchangeably, these words have different meanings. *Fish* is used either as singular noun or to describe a group of specimens from a single species. *Fishes*describes a group of different species:

> *"The present encyclopaedic work on fishes is designed to impart latest information in the light of recent developments and researches in the field besides the basic conventional information.*

This text is designed to approach the morphology, anatomy, physiology and development of the Fishes in a coherent way since it is most natural for a student in studying the architectural elements of an animal body in the quest to know how they function.

Profusely illustrated, this work features a text, reference and a laboratory guide. It deals with the structures and physiological phenomena of each system. Origin, Adaptative radiation and other general topics are also dealt in details. No doubt, this work, in three volumes, will not only meet the requirement of Indian students but will also be useful as a guid.

Adaptations in Fishes

Missouri fishes inhabit many different aquatic habitats from riffles and pools of large rivers and small streams to natural and artificial lakes, ponds and oxbows. Each species has adapted to survive in one or more of these habitats, but few can survive in all habitats.

Traits that allow a species to live in a particular habitat can be adaptations in temperature tolerance, salt tolerance, the ability to breath air when needed, and in body form.

Adaptations of a fish's body that allow a species to survive in different habitats include shape of the body, fin location and mouth size and orientation. In general, body forms of Missouri fishes can be grouped into seven different categories: rover predator, lie-in-wait predator, deep-bodied fish, eel-like fish, bottom rover, bottom clinger and surface-oriented fish.

Rover Predator

Temperate and black basses (like the largemouth bass illustrated above), trout and walleye are examples of fish with body shapes adapted to a rover-predator lifestyle. Their bodies are elongate, their fins are distributed around the body evenly, their head is pointed with a relatively large mouth at the tip, and they usually have a forked tail, an indicator of a fast swimmer. This type of fish roves through several types of habitats from still to swift waters, and uses speed to catch prey. Many of our favourite Missouri game fish are rover predators.

Lie-in-wait predator

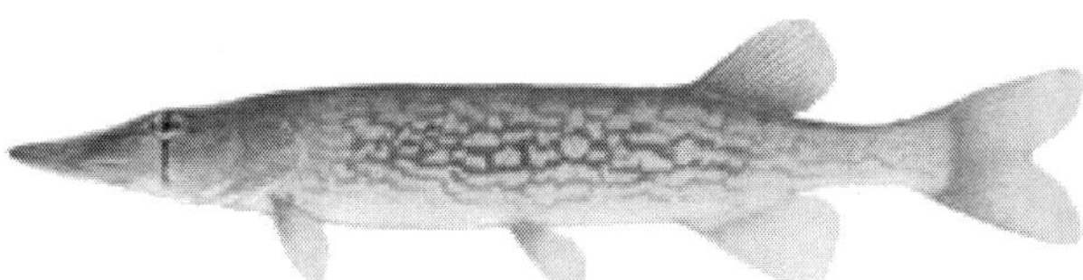

Pickerel (illustrated above) and gar are Missouri species with the lie-in-wait predator body adaptation. Their bodies are elongate, their fins are moved back on the body to provide thrust for chasing prey, their head is pointed and flattened on top, and they have a mouth full of teeth at the tip. This type of fish will lie hidden by cover in still or slow moving waters and dart out to snap up prey.

Deep-bodied Fish

Many sunfish (like the bluegill illustrated above) and buffalofish species exhibit the deep-bodied form. Deep-bodied fishes are flattened laterally and have long dorsal and anal fins.

Their pectoral and pelvic fins are located high on the body. This type of fish mainly occurs in still waters and are adapted for maneuverability in heavy cover. They typically feed from the bottom or pick slow-moving prey from the water column.

Eel-like Fish

American eel (illustrated above) and lampreys are examples of Missouri eel-like fish. They have very elongate bodies with reduced pelvic and pectoral fins, very long anal and dorsal fins, and unforked tails. This type of fish occurs in still to moderately swift waters and is adapted for maneuverability in tight places and burrowing in soft substrates. Eels typically sneak up on prey, and lampreys filter food from the bottom with some species becoming parasitic as adults.

Bottom Clinger

Sculpins (like the banded sculpin illustrated above) and many darter species are adapted as bottom clingers. They tend to have flattened heads and large pectoral fins that are angled to keep the fish on the bottom in swift currents. This type of fish occurs in the swift water of riffles where they pick invertebrates from the rocks

Bottom Rover

Many Missouri catfish and suckers (like the golden redhorse illustrated above) are adapted as bottom rovers. These species have a body shape similar to the rover predator except the back is humped, the head is flattened, and the pectoral fins are enlarged. This type of fish occurs in still to swift waters and uses speed to avoid predators. The mouths of bottom rovers are placed in several different positions depending on their feeding style. Omnivours like catfish have a terminal mouth that allows them to take prey if the opportunity arises, generalist bottom feeders like many suckers have a subterminal mouth, and bottom feeders like carp have ventral mouths with protrusible lips for sucking ooze. Many bottom rovers also have well developed barbels for locating food by feel.

Surface-oriented

Missouri's topminnows (like the blackstripe topminnow illustrated above) and studfish are surface-oriented species. They are typically small-bodied and have a flat head with upturned mouth. Their dorsal fin is moved back on their body. This type of fish occurs in still waters and feeds on organisms that fall in the water. Many surface-oriented species can survive in deoxygenated waters by breathing the surface layer of water that always contains some oxygen.

How Fish Use Adaptations

Colouration

Fish display a wide variety of colours and colour patterns. Skin colouration can have many functions. Many fish have colour patterns that help them blend in with their environment. This may allow the fish to avoid being seen by a predator. Some fish, such as the flat fishes (Pleuronectiformes), can change their skin colouration to match the surrounding habitat.

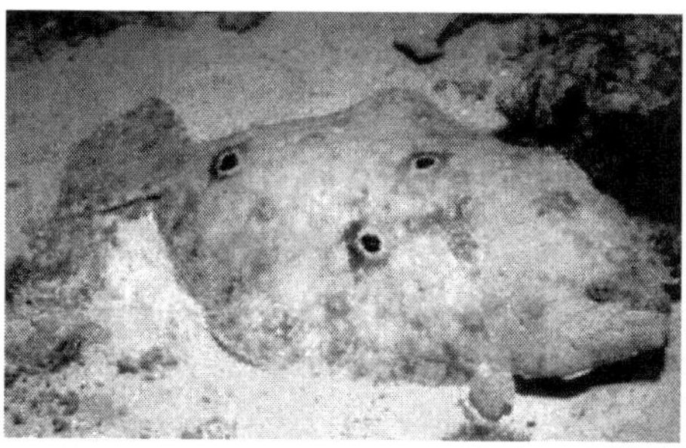

Figure: Gulf flounder

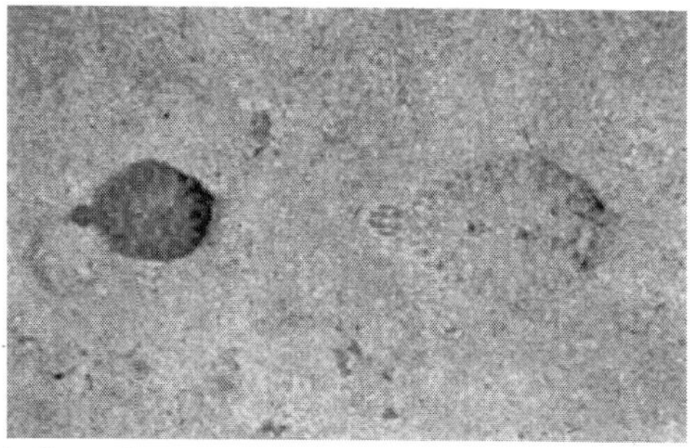

Figure: Wide-eyed flounder.

Fish can also have disruptive markings to hide body parts. Species such as the jackknife fish (*Equetus lanceolatus*), high-hat (*Equetus acuminatus*) and some angel fishes (Pomacanthidae), have dark lines that run through the eyes. These lines may serve to hide the eyes so that other animals can not tell where the fish is looking or even if it is a fish. Also, horizontal lines may be a sight-line for aiming attacks on prey. Some fishes, like butterflyfishes (Chaetodontidae), have spots on their body that resemble eyes. This may serve to confuse prey and predators alike. In addition to colouration, some fish, like the sea dragon (*Phyllopteryx*), have body shapes that can further mimic their habitat.

Figure: Foureye butterflyfish.

Figure: High-hat, family Sciaenidae *David Snyder*

Fish colouration can also be useful in catching prey. Many sharks exhibit colouration known as counter shading. Sharks that have counter shading are dark on the dorsal (upper) side and light on the ventral (lower) side. With this colour scheme any prey looking down on the shark will see a dark shark against a dark sea bottom, making it hard to detect the shark. Conversely, any prey looking up at the shark, will see the light belly of the shark on the light background of the ocean surface water lit by the sun or moon.

Figure: Gray reef shark.

Figure: Silvertip shark.

Colouration can also be used to advertise. Fishes like the darters (Percidae) and sticklebacks (*Gasterosteus*), may use colour to attract and recognise potential mates.

Figure: Rainbow darter.

Figure: Coosa darter.

Light Organs

Some marine fish have the ability to produce light through bioluminescence. Most light-producing fish live in mid-water or are bottom-dwelling deep sea species. In fish, bioluminescence can occur two different ways: through symbiotic bacteria living on the fish or through self-luminous cells called photophores. Some species of deep sea angler fish (Lophiiformes) may use this light to attract prey, while others, like the Atlantic midshipman (*Porichthys plectrodon*), may use this light to attract mates.

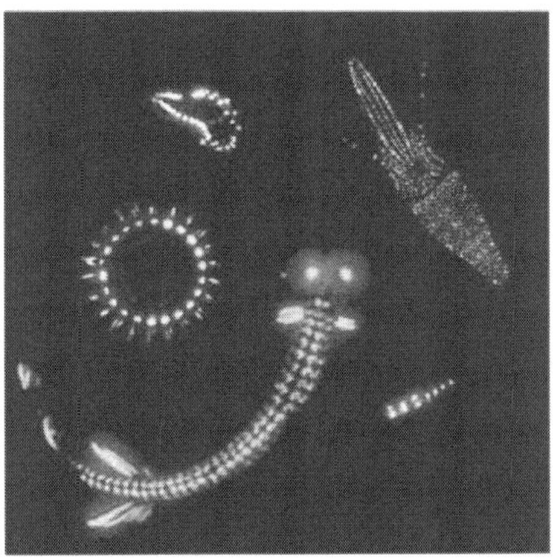

Figure: Deep water bioluminescent organisms *courtesy NOAA*

Venom

Many fish may use venom as a form of defence. Most venomous fish deliver the toxins through the use of a spine. Venomous spines are found in a wide variety of fish including stingrays, chimaeras, scorpionfishes, catfishes, toadfishes, rabbit fishes, and stargazers. Venomous spines can have poison glands along the grove of the spine, as with stingrays, or at the base of the spine, as in some catfish. While humans can be stung by a multitude of fishes, few species are life-threatening.

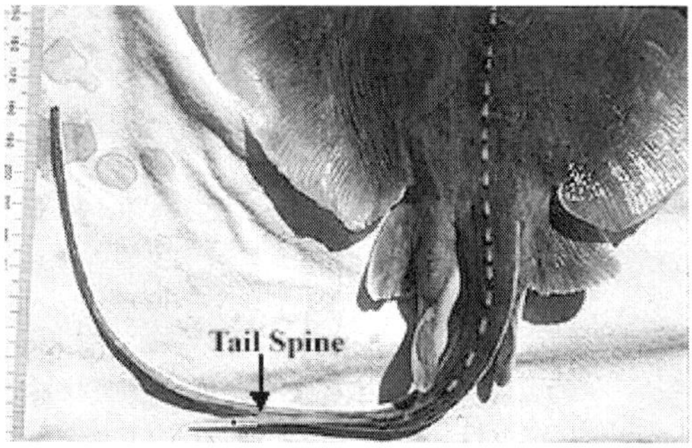

Figure: Stingray tail spine.

Electric Organs

Elasmobranchs (sharks, skates, and rays) possess an electric sense system known as the ampullae of Lorenzini. This system consists of many tiny gel-filled canals positioned on the head of the fish. Through this system these fishes are able to detect the weak electric fields produced by prey. It is also believed that these fish can use this sense to detect the electric fields they induce when swimming through the earth's magnetic field, as a sort of compass. Since the fishes are able to generate the fields they detect, this is a form of active electro-orientation. Some species of skates and rays also have electricity-producing organs. The electric rays have paired electric organs located on either side of the head, behind the eyes. With these organs, electric rays are able to shock and stun their prey. The skate's electric organs are located near the tail. However, these electric organs only produce weak electric fields not capable of stunning prey. Researchers believe that the skate's electric organs are used for communication and mate location.

Figure: Marbled electric ray.

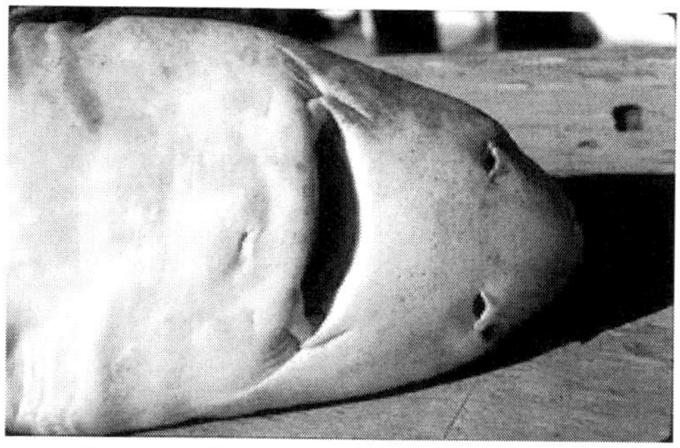

Figure: Roughskin spiny dogfish.

The electric eel can also produce electric fields. These eels use weak electric fields for navigation, prey location, and communication. Additionally, these eels can produce strong electric fields to stun potential prey. The strength of the "shock" is related to the size of the eel, with larger individuals being able to produce more of a "shock."

Figure: Electric eel.

Origin and Evolution of Fishes

The History of Animal Evolution

For many people animals are perhaps the most familiar, and most interesting, of living things. This may be because we are animals ourselves. As such, we have a number of features in common with all the organisms placed in the animal kingdom, and these common features indicate that we have a shared evolutionary history.

All animals and plants are classified as multicellular eukaryotes: their bodies are made up of large numbers of cells, and microscopic inspection of these cells reveals that they contain a nucleus and a number of other organelles. Compared to prokaryotic organisms such as bacteria, plants and animals have a relatively recent evolutionary origin. DNA evidence suggests that the first eukaryotes evolved from prokaryotes, between 2500 and 1000 million years ago. That is, eukaryotes as a taxon date from the Proterozoic Era, the final Era of the Precambrian. Fossils of both simple unicellular and more complex multicellular organisms are found in abundance in rocks from this period of time. In fact, the name "Proterozoic" means "early life".

Plants and animals both owe their origins to endosymbiosis, a process where one cell ingests another, but for some reason then fails to digest it. The evidence for this lies in the way their cells function. Both plant and animal rely on structures called mitochondria to release energy in their cells, using aerobic respiration to produce the energy-carrying molecule ATP.

There is considerable evidence that mitochondria evolved from free-living aerobic bacteria: they are the size of bacterial cells; they divide independently of the cell by binary fission; they have their own genome in the form of a single circular DNA molecule; their ribosomes are more similar to those of bacteria than to the ribosomes found in the eukaryote cell's cytoplasm; and like chloroplasts they are enclosed by a double membrane as would be expected if they derived from bacterial cells engulfed by another cell.

Like the plants, animals evolved in the sea. And that is where they remained for at least 600 million years. This is because, in the absence of a protective ozone layer, the land was bathed in lethal levels of UV radiation. Once photosynthesis had raised atmospheric oxygen levels high enough, the ozone layer formed, meaning that it was then possible for living things to venture onto the land.

The oldest fossil evidence of multicellular animals, or metazoans, is burrows that appear to have been made by smooth, wormlike organisms. Such trace fossils have been found in rocks from China, Canada, and India, but they tell us little about the animals that made them apart from their basic shape.

- The Ediacaran animals
- The Cambrian "explosion" and the Burgess Shale
- What caused the Cambrian "explosion"?
- A foot on the land
- The earliest vertebrates
- Appearance of the fish
- The jawless fish.

The Ediacaran Animals

Between 620 and 550 million years ago (during the Vendian Period) relatively large, complex, soft-bodied multicellular animals appear in the fossil record for the first time. While found in several localities around the world, this particular group of animals is generally known as the Ediacaran fauna, after the site in Australia where they were first discovered.

The Ediacaran animals are puzzling in that there is little or no evidence of any skeletal hard parts i.e. they were soft-bodied organisms, and while some of them may have belonged to groups that survive today others don't seem to bear any relationship to animals we know. Although many of the Ediacaran organisms have been compared to modern-day jellyfish or worms, they have also been described as resembling a mattress, with tough outer walls around fluid-filled internal cavities - rather like a sponge.

As a group, Ediacaran animals had a flat, quilted appearance and many showed radial symmetry. They ranged in size form 1cm to >1m, and have been classified into three main groups on the basis of their shape: discoidal, frond-like, or ovate-elongate. The large variety of Ediacaran animals is significant, as it suggests there must have been a lengthy period of evolution prior to their first appearance in the fossil record.

The Cambrian "Explosion" and the Burgess Shale

The Ediacaran animals disappear from the fossil record at the end of the Vendian (544 million years ago). In their place we find representatives of almost all the modern phyla recognised today: sponges, jellyfish and corals, flatworms, molluscs, annelid worms, insects, echinoderms and chordates, plus many "lesser" phyla such as nemertean worms. These "modern" organisms appear relatively quickly in the geological time scale, and their abrupt appearance is often described as the "Cambrian explosion" however, bear in mind that the fossil record of the "explosion" is spread over about 30 million years. I keep taking things out of brackets because it is interesting relevant and memorable

One of the most famous assemblages of Cambrian fossils comes from the Burgess Shale of British Colombia. The rocks of the Burgess Shale were laid down in the middle Cambrian, when the "explosion" had already been underway for several million years. They contain familiar animals such as trilobites, molluscs and echinoderms, but also the first appearance of brachiopods, and some odd animals, e.g. *Opabinia*, that may have belonged to extinct phyla. Even an early chordate, *Pikaia*, has been found in this fossil assemblage.

The Burgess Shale fossils are important, not only for their evidence of early variety among animal forms, but also because both soft parts of animals and their hard bodies (i.e. the whole animal) is preserved, and animals that were entirely soft-bodied. Preservation of soft-bodied organisms is rare, and in this case seems to have occurred when the animals were rapidly buried in a mudslide down into deep, anaerobic waters, where there was little bacterial decay. Prior to the discovery of this fossil assemblage, early in the 20th century, there was no evidence of soft-bodied animals from the Cambrian (remember that this is before the Ediacaran fauna were found).

These fossils also provide good evidence of predatory animals (e.g. Anomalocaris), and therefore of complex predator-prey relationships. They also give insights into how evolution might have progressed relatively early in the history of multicellular animals, and in fact some authors view the Cambrian as a period of extreme “experimentation” and diversity.

What Caused the Cambrian “Explosion”?

The cause of the proliferation of animal forms in the Cambrian is a matter of considerable debate among scientists. Some point to the increase in atmospheric oxygen levels that began around 2000 million years ago, supporting a higher metabolic rate and allowing the evolution of larger organisms and more complex body structures. Changed ocean chemistry would have played a part here, allowing for the first time the development of hard body parts such as teeth and supporting skeletons based on calcium carbonate ($CaCO_3$), and also supporting higher levels of primary production as a result of increased concentrations of phosphates and nitrates. The mass extinction that marked the end of the Vendian period would have opened up ecological niches that the new animals exploited, as would habitat changes wrought by continental drift.

Genetic factors were also crucial. Recent research suggests that the period prior to the Cambrian explosion saw the gradual evolution of a “genetic tool kit” of genes (the homeobox or ”hox” genes) that govern developmental processes. Once assembled, this genetic tool kit enabled an unprecedented period of evolutionary experimentation and competition. Many forms seen in the fossil record of the Cambrian disappeared without trace. Future evolutionary change was then limited to acting on the body plans that remained in existence.

Recently many scientists have begun to question whether the Cambrian explosion was a real event, or a reflection of the patchiness of this ancient fossil record. Genetic data suggest that multicellular animals evolved around 1000 million years ago; this is supported by fossil embryos from rocks in China that date back 600 million years. These embryos are more complex than those of simple organisms such as sponges and jellyfish, which suggests that multicellular animals must have evolved much further back in time. In addition, trilobites were a very diverse group even early in the Cambrian, and some scientists suggest that this indicates that the arthropod group must have had a much earlier evolutionary origin.

A Foot on the Land

Whatever their origins, animals may have ventured onto land early in the Cambrian. Previously scientists believed that animals did not begin to colonise the land until the Silurian (440 - 410 million years ago). However, the 2002 discovery of the footprints of animals that scuttled about on sand dunes about 530 million years ago has changed this view. These animals were arthropods, and resembled centipedes about the size of crayfish. They probably didn't live on land, instead coming ashore to mate or evade predators. At this time the only land plants appear to have resembled mosses .

The Earliest Vertebrates

Animals continued to diversify in the Ordovician seas (505 - 440 million years ago). They were mostly invertebrates, including graptolites , which were stick-like branching colonies of tiny animals, together with brachiopods, trilobites, cephalopods, corals, crinoids and conodonts. We now place the conodonts with the chordates, but for a long time they were known only by their tiny, but very common, teeth.

In terms of number of species invertebrates were by far the most common Ordovician animals - as they still are today. However, members of another taxon were also evolving in the Ordovician seas. These were the fish.

Appearance of the Fish

Like the conodonts, fish are members of the chordate phylum because they display certain defining characteristics: a dorsal stiffening rod called the notochord, a dorsal nerve cord, pharyngeal gill slits and a tail that extends beyond the anus. However, fish are placed in the subphylum Vertebrata , because they also show the development of skeletal features such as a backbone, skull, and limb bones. Not all the modern groups of fish were represented in the Ordovician oceans. At this time only the jawless fish had evolved from a chordate ancestor. The sharks and their relatives and two extinct groups, the placoderms (which had bony plates covering their heads) and the acanthodians (the first known jawed vertebrates, with a skeleton of cartilage) made their appearance in the Silurian. However, neither the sharks nor the agnathans became common until the Devonian. The other two living lineages, the ray-finned (e.g. carp and kahawai) and the lobe-finned fish (e.g. lungfish and the coelacanth), evolved during the Devonian period.

The Jawless Fish

Agnathans, or jawless fish, were the earliest fish: an excellent fossil of *Haikouichthys ercaicunensis* dates back about 530 million years, to the Cambrian. Previously the earliest-known agnathans were dated to around 480 million years ago. Agnathans have traditionally been placed with the vertebrates due to the presence of a skull, although the modern forms such as hagfish lack a vertebral column. The earliest agnathans were Ostracoderms. They were bottom-feeders and were almost entirely covered in armour plates. When the sharks and bony fish began to evolve, around 450 million years ago, most ostracoderms became extinct. Only the lineage that produced the modern hagfish and lampreys survived.

Fish Evolution

Evolution

The first fishes, and indeed the first vertebrates, were the ostracoderms, which appeared in the Cambrian Period, about 510 million years ago, and became extinct at the end of the Devonian, about 350 million years ago. Ostracoderms were jawless fishes found mainly in fresh water. They were covered with a bony armour or scales and were often less than 30 cm (1 ft) long. The ostracoderms are placed in the class Agnatha along with the living jawless fishes, the lampreys and hagfishes, which are believed to be descended from the ostracoderms.

The first fishes with jaws, the acanthodians, or spiny sharks, appeared in the late Silurian, about 410 million years ago, and became extinct before the end of the Permian, about 250 million years ago. Acanthodians were generally small sharklike fishes varying from toothless filter-feeders to toothed predators. They are often classified as an order of the class Placodermi, another group of primitive fishes, but recent authorities tend to place the acanthodians in a class by themselves (class Acanthodii) or even within the class of modern bony fishes, the Osteichthyes. It is commonly believed that the acanthodians and the modern bony fishes are related and that either the acanthodians gave rise to the modern bony fishes or that both groups share a common ancestor. The placoderms, another group of jawed fishes, appeared at the beginning of the Devonian, about 395 million years ago, and became extinct at the end of the Devonian or the beginning of the Mississippian (Carboniferous), about 345 million years ago. Placoderms were typically small, flattened bottom-dwellers. The upper jaw was firmly fused to the skull, but there was a hinge joint between the skull and the bony plating of the trunk region.

The cartilaginous-skeleton sharks and rays, class Chondrichthyes, which appeared about 370 million years ago in the middle Devonian, are generally believed to be descended from the bony-skeleton placoderms. The cartilaginous skeletons are considered to be a later development. The modern bony fishes, class Osteichthyes, appeared in the late Silurian or early Devonian, about 395 million years ago. The early forms were freshwater fishes, for no fossil remains of modern bony fishes have been found in marine deposits older than Triassic time, about 230 million years ago. The Osteichthyes may have arisen from the acanthodians. A subclass of the Osteichthyes, the ray-finned fishes (subclass Actinopterygii), became and have remained the dominant group of fishes throughout the world. It was not the ray-finned fishes, however, that led to the evolution of the land vertebrates.

The ancestors of the land vertebrates are found among another group of bony fishes called the Choanichthyes or Sarcopterygii. Choanate fishes are characterised by internal nostrils, fleshy fins called lobe fins, and cosmoid scales. The choanate fishes appeared in the late Silurian or early Devonian, more than 390 million years ago, and possibly arose from the acanthodians. The choanate fishes include a group known as the Crossopterygii, which has one living representative, the coelacanth Latimeria. During the Devonian Period some crossopterygian fishes of the order (or suborder) Rhipidistia crawled out of the water to become the first amphibians.

Classification

The classification of fishes is a subject of considerable debate. The living fishes are often divided into three different classes. Divisions within these classes, however particularly within the bony fishes (Osteichthyes) are much in dispute, with different names being applied to the same group and with a given group being regarded as either a subclass, order, or some other rank, depending upon the authority consulted.

The following is a general classification; to help clarify terms that may be found in other sources, it also includes as many additional definitions as space allows.

Class Agnatha or Cephalaspidomorphi, the jawless fishes.

Subclass (or order) Cyclostomata, the lampreys and hagfishes.

(In certain classifications, the lampreys and hagfishes are each considered separate superclasses: Cephalaspidomorphi and Pteraspidomorphi, respectively.)

Class Chondrichthyes, the cartilaginous-skeleton fishes.

Subclass Holocephali, the chimaeras, or ratfishes.

Subclass Elasmobranchii, the sharks, skates, and rays.

Class Osteichthyes, the bony fishes.

Subclass (or order) Crossopterygii, the coelacanth.

Subclass (or order) Dipnoi or Dipneusti, the lungfishes.

(In some classifications, the above two subclasses are treated as orders of a single subclass, the Choanichthyes or Sarcopterygii, the lobe-finned fishes.)

Subclass Actinopterygii, the ray-finned fishes.

Infraclass (or superorder) Chondrostei, the primitive ray-finned bony fishes: sturgeons, paddlefish, and bichirs.

(In some classifications, the bichirs are placed in a subclass of their own, the Brachiopterygii.)

Infraclass (or superorder) Holostei or Neopterygii, the intermediate ray-finned fishes: gars and the bowfin.

(In certain classifications, the gars are treated as a separate superorder, the Ginglymodi. The term Ginglymodi also has been used to designate the gars as an order, but this term has been replaced at the ordinal level by the term Lepisosteiformes; orders are now indicated by the ending -formes.)

Infraclass (or superorder) Teleostei or Neopterygii, the advanced bony fishes: herring, salmon, perch.

The Evolutionary Steps of Fish

Fish are incredible diverse group, made up of three living classes that hold an important place both in modern ecology and in evolutionary history. Fish are incredibly successful, they are numerous and have a huge variety of adaptations. They are successful in every water filled habitat. Furthermore, fish are the first known vertebrates and also the stepping stone to all land-walking vertebrates (tetrapods). Fish are complicated, diverse, and have a long evolutionary history.

In order to understand the place of fish as both a transitory group and as a modern one it is important to understand the times in which and from which they evolved. Fish are the first known true chordates. The first vertebrate that has been found is the Upper Cambrian fossil Anaspis, which is more than 500 million years old. This fossil, while being fragmentary, is thought to be an armoured, jawless fish. Fish did most of their evolving between five million and three and a half million years ago. These two periods were known as the Silurian and the Devonian periods. In yhe middle Silurian, the jawless fishes had diversified, but it was not until the Devonian that the true variety of fishes really flourished. In fact, the Devonian, is often referred to as the "Age of Fishes". Towards the end of the Devonian the first tetrapods (vertebrates which evolved true legs with which they could walk on land) had evolved from one specific branch of fish. Fish greatly specialised in their aquatic niche during both the Devonian and the Silurian and part of this evolution led to adaptations to land in the form of amphibians.

Fish are in no way simply a stepping stone to amphibians, they are a much more significant than this. They have evolved to be masters of their domain, the water. They come in many forms, have the ability to eat a huge variety of foods, and have populated almost every body of water. In fact, fish are the most common vertebrate, with there being approximately 24,000 species alive today. This number is mind boggling when put in perspective; the next most common vertebrate are birds with a mere 8,600 species. This multitude of species ranges in size, morphology, agility, and adaptations to environment. Fish have been broken down into a series of classes that separate them based on characteristics. These groups help to classify the wide variety of species that make up fish and help lead to understand of the current and evolutionary niche of fish.

The first fish to evolve were the Agnathans (Class agnatha). These jawless fishes are the first vertebrates. These fish have round mouth parts that could be used for sucking or filter feeding. These rasping, sucking mouths are currently found on modern lampreys and hagfishes. These fish were often extremely armoured in order to help them protect themselves. One group that evolved before the Silurian were the Ostracoderms which have been described as "small, blunt-headed forms they fed on debris in the mud, bullet shaped swimmers, and some with an astonishing array of spines and crests on their heads." Most of these types of fish are currently extinct with the exceptions of the lampreys and the hagfish.

From these bottom feeding, jawless fish came the evolution of jawed fish. Jaws evolved only once (rather than evolving multiple times in different species through parallel evolution). Jaws evolved from gill arches which are the bony parts between gill slits. It is thought that a gill arch in an agnathan became fused to its skull. The upper part of the gill support became the top jaw and the bottom part of the gill support became the bottom jaw. Embryology points to this and the arrangement of nerves in shark heads and most simple fishes shows that jaws are in line with gill arches. While fish had the first bony jaws, they also have some of the most complicated. While the human head has only one moving part (the jaw), the head of a fish may have more than twenty-four bones that may move together in feeding. The evolution of the jaw is incredibly important because it led to fish to be able to ingest a much wider variety of foods and allowed them to be active hunters as opposed to passive filter feeder. This led to a wide variety of adaptations in morphology. Fish became more agile to be better predators, they were able to reduce their armour because they were less vulnerable, and their muscle density was able to decrease becuase they no longer led such a sluggish lifestyle.

The first jawless fishes include the acnathodians and placoderms. Even early in their development the jawless fishes showed great diversity. The acnathodians appeared around four and half million years ago. They were the first jawed fishes to evolve and they tended to be small, streamlined, and had huge eyes. Placoderms on the other hand appeared in the early Silurian and dominated the Devonian. These were some strange looking fish! As described by Peter Gore, "the Placoderms had bony armour that covered the head and forepart of the body. In many, a movable joint between the head and body armour let the head rock back to open the mouth wide. The primitive jaws had jagged bony edges that served as teeth. The tail end usually lacked protection". At the beginning of the Devonian, Placoderms were small, but they soon increased in size. The Dunkleosteus, a placoderm, grew to be as large as 35 feet, had well developed jaws, with fang like teeth. The front of the trunk was heavily armoured and the hind part was either bare or covered with small scales. The Dunkleosteus and all the other large placoderms are extinct, but in the Devonian they dominated both salt and freshwater.

While the huge, menacing placoderms have long since vanished from the earth, members of the Class Chondritchthyes, some of which are equally menacing, have not. This class includes the sharks, skates, and rays (along with some other fish) and these first evolved between 400 and 450 million years ago. They probably shared a common ancestor with the placoderms. This class is commonly refer to as the cartilaginous fish because they lack true bone, instead they have cartilage and calcified cartilage for internal support. This type of skeleton is extremely light and flexible and helps these species be agile predators. Chondrithyes have evolved two kinds of jaws; sharks have extremely powerful biting and crushing jaws while skates and ray uses their jaws to be bottom feeders searching for mollusks. Sharks are often referred to as living fossils since they have not done much evolving since the Late Paleozoic. This is not to say that the sharks that exist now are they only sharks ever to exist. After the decline of the placoderms, there was a shark explosion that led to sharks with mouth's filled with coils of serrated teeth, or bony protuberances on their fins. These Paleozoic sharks could be a large as 14 m, more than double the size of the current Great White Sharks.

In terms of currently living fish, Class Chondrichtyes make up only 900 species. The bulk of the fish are from Class Osteichtyes, the true bony fish. There are a reported 19,000 bony fish (this is a very conservative estimate). This is an incredibly diverse class which includes everything from tuna to eels. These fish evolved more that 410 million years ago (Late Silurian), but did not rise until the middle Devonian as the placoderms and huge sharks began to recede in dominance. The bony fish, while being varied, all share an extremely important characteristic: a swim bladder. This probably evolved from lungs which had appeared in some freshwater species. The swim bladder is a internal structure which allows bony fish to float easily at any water level. The bony fish can be broken up into three major groups. The first are the ray-fins. These are most of the common bony fish such as tuna, bass, salmon. This group is very diverse and inhabits both fresh and salt water. The second group are the lungfish which are freshwater fish. The three genera which still exist are bottom feeders who use jaws to crush their prey. They can get up to 90% of their oxygen from the air if necessary. This can be very important when these fish live in areas prone to drought. The third group are the lobed-fin fish. This entire group was though to be extinct until the 1930s when a live one was recovered. It was called a coelacanth and is described as having a incredible swimming style which "consists of walking-like movements of the various paired fins, with some wriggling of the body but no side to side beating of the tail as in most other fishes." While this species is extremely important as a living fossil, the rhipidsistains, a group of lobe-fins, are also key sine they are the source of the first land vertebrates.

While there is still much controversy about the exact steps of evolution from lobe-finned fish to walking, land living, vertebrates (in the form of amphibians). There was originally much controversy over whether the limbs of tetrapods (limbed vertebrates) evolved in water or on land as an emergency feature. The original hypothesis was put out by Alfred Romer in the 1940s. It was that early lung fishes in times of drought had used their fleshy fins to pull themselves from a pond that was trying onto land to search for a more fruitful water source. The fish that made this journey successfully were able to reproduce and their offspring began to have modified limbs which were able to allow them to move from water source to water source and eventually these developed into true limbs. Romer relied on the idea that the end of the Devonian was very harsh and that fish would need to adapt rapidly in order to survive, but many believe that this is not the case. In fact, another theory is that the Devonian was so conducive to fish that the were able to have many offspring survive and that these were drawn to the land with its not yet exploited food sources. This theory still supports Romer's ideas that limbs evolved to aid in walking on land. A newer theory as developed by Jenny Clack in the early 1990s due to her studying of Ancathostega fossils. Ancathostega is one of the earliest tetrapods found, and by far the most complete. These creatures had four limbs, but its limbs were still proportioned like fins (very long radius in comparison to ulna) and had eight fingers and a weak wrist. They also breathed like fish, and had ribs that were too small to hold up its gut. In other words, these tetrapods were not adapted to walk on land. They had many adaptation to water, including a powerful tail that would lead to agility. Clack believes that Acanthostega evolved limbs to help maneuver through the Devonian habitat which consisted of very dense wetlands. These creatures could maneuver through these dense branches and plants by using their limbs rather than just having to "wiggle" through. They could also use the limbs to anchor themselves to wait silently for prey and then use their water adaptations to snag prey deftly. These huge tetrapods may have began to use land in a very gradual fashion, for breeding or to escape predators. The evolution of fish into land walking tetrapod is in no way close to being understood, but as more specimens are being found (a very early one was just found near the Poconos) the path of evolution may become more clear.

In conclusion, there is much to be learned from fish. They have evolved to fill every niche that includes water and have an incredible diversity of adaptations. Their evolutionary history is complicated and complex. They began as jawless, bottom feeders and evolved into sharks, rays, tunas, and many extinct species including the placoderms. Fish are also important as an evolutionary step. They are the first known true vertebrates and also provide a very key stepping stone to the evolution of all land living vertebrates. This evolutionary step from bony fish to amphibians is still not fully understood, and the research is very current. The multiple theories all have merit and a crucial debate is whether limbs evolved out of a push to be land dwellers or as an adaptation that made certain fish able to manipulate the water of the Devonian period. Many more specimens need to be found in order for this debate to lead to one side or the other. Fish are too often dismissed as either ruthless killers (sharks) or as a source of food, but in truth they are much more vital to our own history than most people care to think about.

The Evolutionary Origins of Syngnathidae: Pipefishes and Seahorses

The Syngnathidae (seahorses and pipefishes) are a large family of close to 300 marine, brackish and freshwater species (Froese & Pauly, 2010), all of which share an exceptional form of reproduction, male pregnancy. This key feature of the group has made them important model organisms for the study of sexual selection.

In addition to their unique reproductive behaviour, syngnathids are well known for their highly specialised morphology, and the diversity of morphological forms found in this group has complicated efforts to understand their evolutionary origins and pattern of diversification. This exceptional morphological variation is reflected in the current taxonomy of the group: 14 of the 54 currently recognised syngnathid genera are monotypic (Froese & Pauly, 2010), and the majority of genera are composed of fewer than three species.

The family Syngnathidae has traditionally been included as a member of the order Gasterosteiformes, which includes 11 families in two suborders: the Gasterosteoidei, with the Hypoptychidae, Gasterosteidae and Aulorhynchidae; and the Syngnathoidei, with the Indostomidae, Aulostomidae, Fistulariidae, Macroramphosidae, Centriscidae, Pegasidae, Solenostomidae and Syngnathidae. The evolutionary history of the Gasterosteiformes itself has been the subject of controversy since its inception because of the derived and highly reductive morphology of its constituent families. Major monophyletic groups of families within the Gasterosteiformes have long been recognised, including Gasterosteidae and Aulorhynchidae (as superfamily Gasterosteoidea), Aulostomidae and Fistulariidae (Aulostomoidea), Centriscidae and Macroramphosidae (Centriscoidea), and Syngnathidae and Solenostomidae (Syngnathoidea). Studies using morphology have proposed monophyly of the order and suggested sister groups based on weak evidence, and while the close relationships of superfamilies is well supported, the relationships among these family pairs remains unclear. In contrast, molecular evidence clearly refutes the monophyly of the Gasterosteiformes, placing gasterosteoids close to the cottoid–zoarcoid lineage (Imamura & Yabe, 2002), excluding the Indostomidae, and placing both groups distant from syngnathoids. While the sister group of syngnathoids remains unknown and the relationships among syngnathoid lineages are poorly resolved, the family Solenostomidae (ghost pipefishes) has been faithfully recovered as the sister group of the Syngnathidae in both morphological and molecular analyses, an evolutionary relationship key to understanding the evolution of male parental care in the family Syngnathidae.

Higher Level Phylogenetics: Morphology vs. Molecules

Morphological Analyses

Linnaeus (1758) codified the prevailing concept of a wide separation between the currently recognised gasterosteoids and syngnathoids in the 10th edition of *Systema Naturae*. Gasterosteids were placed in the Artedian Thoracici, while syngnathoids were divided into two groups, the Abdominales and the Branchiostegi. *Pegasus* was classified separately from the syngnathoids as a chondropterygian (Gill, 1884). In a major revision as part of the final edition of *Systema Naturae*, Linnaeus (1768) reclassified the syngnathoids *Centriscus, Syngnathus* and *Pegasus* together in the Amphibia Nantes, reflecting his erroneous belief that all these fishes possessed both lungs and gills, a contention which perplexed many of his contemporaries (Pietsch, 1995). The position of *Fistularia* and the gasterosteids remained unchanged in this edition of *Systema Naturae*. Linnaeus' systematic groupings of gasterosteoids and syngnathoids were retained for almost half a century, with only minor changes in taxonomic nomenclature during this time.

As part of his effort to classify fishes into natural groups on the basis of unique characters, Cuvier (1816) placed gasterosteiforms in three widely separated lineages and established two key new groups: the Lophobranches and 'Les Bouches en Flute'. Cuvier's Lophobranches united the pegasids, solenostomids and syngnathids, while 'Les Bouches en Flute' comprised the aulostomoids and centriscoids, marking the first time that these two lineages were closely associated. Cuvier & Valenciennes (1829) tentatively placed the Gasterosteidae in a primarily scorpaeniform group, foreshadowing current ideas of scorpaeniform affinities based on molecular data.

The ideas of two American workers, Cope and Gill, had a broad influence on the thoughts of later researchers in gasterosteiform systematics. In establishing a filing system for use at the National Museum of Natural History, Washington DC, U.S.A. Gill (1872) was influenced in his ideas of the relationships of gasterosteiforms by Cope (1871). Gill's Lophobranchii contained only the Solenostomidae and the Syngnathidae [although *Hippocampus* was placed within its own family, the Hippocampidae, for reasons detailed by Gill (1896)], while his Hemibranchii was composed of the Gasterosteiformes, with presently recognised gasterosteoids and aulostomoids, and the Centrisciformes, containing only centriscoids. Noting that a linear series of names is inadequate to express genetic relationships, he outlined the evolutionary lineage of fishes and depicted the Lophobranchii arising from the Hemibranchii. These ideas of gasterosteiform evolution followed Cope (1871), except for the placement of the Pegasidae, which was not included. Gill (1884) published a detailed analysis of the Hemibranchii, which contained all gasterosteiforms except syngnathids and solenostomids and argued that the Aulorhynchidae and Gasterosteidae were only slightly differentiated, even going as far as suggesting that they might appropriately be united into one family.

The landmark studies of Jungersen (1908, 1910) provided the first detailed accounts of the morphology of the Centriscidae and Macroramphosidae in 1908 and the Aulostomidae, Fistulariidae, Syngnathidae and Solenostomidae in 1910. In the first of these papers, he concluded that the Solenichthys (the Syngnathoidei less Pegasidae) was a natural group and that the Hemibranchii (Gasterosteoidei) and Pegasidae were not closely related to this group. He argued that several morphological characters were 'found united in the whole community of forms' (his Solenichthys), placing centriscoids and aulostomoids together as a natural group and solenostomids and syngnathids as a separate but closely related lineage. He also concluded that gasterosteoids and pegasids belonged within scorpaeniforms, and were not closely related to syngnathoids, based on investigations that were never published (Jungersen, 1908, 1915). Regan (1909) was heavily influenced by the work of Jungersen (1908) and recognised monophyletic groups comparable to the Aulostomoidea, Centriscoidea and Syngnathoidea, though he did not specify their relationships to one another. In later work, Regan (1913) suggested that gasterosteids were closely related to scorpaeniform fishes, reiterating the earlier conclusions of Cuvier & Valenciennes (1829).

In his influential classification, Berg (1940) listed the Gasterosteiformes in three separate orders: the Gasterosteiformes, Syngnathiformes and Pegasiformes. The Gasterosteiformes contained the gasterosteoids as well as the Indostomidae, although as*incertae sedis* [the phylogenetic affinities of this family have remained controversial since its initial description (Prashad & Mukerji, 1929)]. The Syngnathiformes included all syngnathoids, except the Pegasidae (which was placed in a separate order at the end of the classification), divided into two suborders following Jungersen (1910): the Aulostomoidei, with aulostomoids and centriscoids, and the Syngnathoidei, with solenostomids and syngnathids. Perhaps the most important higher level work in pre-cladistic systematic ichthyology was the classification of Greenwood *et al.* (1966), which summarised the state of knowledge of the evolutionary relationships of the main phyletic trends of teleosts. Following Berg's ideas of gasterosteiform relationships, Greenwood *et al.* (1966) tentatively brought all families together into a united Gasterosteiformes, indicating that the relationships of this group were 'still to be assessed'.

Although a description of the cladistic method, which emphasizes the pre-eminence of shared evolutionary novelties (*i.e.* synapomorphy) in revealing evolutionary relationships, was first published by Hennig (1950), the influence of this methodology was not felt in ichthyology until the English translation of his work became available in the U.S.A. (Hennig, 1966). The change in systematic rigour with the introduction of cladistic techniques was immense, as the methodology was applied to systematic studies of many taxa of fishes in the early 1970s. In the first cladistic treatment of the Gasterosteiformes, Pietsch (1978) firmly supported an expanded Gasterosteiformes, including the Pegasidae and adding the Hypoptychidae (the Indostomidae was not formally included in this analysis). Johnson & Patterson (1993) added new characters to corroborate Pietsch's (1978) hypotheses of a monophyletic Gasterosteiformes, presented evidence to support monophyly of the Syngnatha (Pegasidae, Solenostomidae and Syngnathidae) and suggested that the Indostomidae should be placed within this group. The Gasterosteiformes were placed into a large group (Smegmamorpha), which included the Synbranchoidei, Mastacembeloidei, *Elassoma*, Mugilomorpha and Atherinomorpha, a relationship which has been called into question in subsequent studies (Springer & Orrell, 2004).

Orr (1995) conducted a phylogenetic analysis of the Gasterosteiformes, primarily based on osteology, and examined the limits, monophyly and relationships of the 11 families of the order. His results supported the monophyly of the Gasterosteiformes and reinforced the monophyly of the traditional major family pairs within the order: the Gasterosteoidea, Aulostomoidea, Macroramphosoidea [=Centriscoidea of Keivany & Nelson (2006)] and Syngnathoidea. The Pegasidae was found to be the sister group of the terminal clade containing the Solenostomidae and Syngnathidae. Orr (1995) excluded *Indostomus* from the Gasterosteiformes and suggested a sister group relationship between the Hypoptychidae and the order Gasterosteiformes.

Understanding the limits and constituents of the taxonomic groups under investigation is critical to a phylogenetic analysis. The influence of taxonomic sampling can be seen in the morphological analysis of Keivany & Nelson (2006), who compiled data of their own with the analyses of several previous morphological analyses in an analysis of all families recently considered to be members of the Gasterosteiformes. They assumed monophyly of the group (Johnson & Patterson, 1993; Orr, 1995; Britz & Johnson, 2002) and used the smegmamorphs of Johnson & Patterson (1993) as outgroups. Their resulting tree placed *Indostomus* as a basal syngnathoid followed by the Pegasidae, which is sister of the Syngnathoidea, the sister of the Aulostomoidea and Centriscoidea [=Macroramphosoidea of Orr (1995)].

Differences among recent morphological analyses of the group demonstrate the instability of the current understanding of gasterosteiform evolution. Orr's (1995) and Keivany & Nelson's (2006) most parsimonious phylogenetic trees were weakest at the nodes of most interest in syngnathoid evolution, the position of the Pegasidae and relationship of the Aulostomoidea and Centriscidae. Outgroup choice and the representation of in-group taxa may have strong effects on the resolution of relationships within a group of interest. With few exceptions, differences in taxonomic representation among morphological and molecular analyses have produced incongruous results in studies of gasterosteiform families.

Molecular Analyses

Two recurring results have been found in all higher level molecular studies published to date: the paraphyly of the Gasterosteiformes and the placement of the Gasterosteoidei together with the cottoid–zoarcoid lineage of Imamura & Yabe (2002). As most higher level molecular studies have had a broad taxonomic focus, taxon sampling of traditional gasterosteiform members has been incomplete, making it difficult to evaluate the relationships of the group as a whole. Early analyses based on complete mitochondrial DNA genome sequences of gasterosteids, *Hypoptychus* and *Indostomus* consistently placed gasterosteids and *Hypoptychus* basal to the cottoid–zoarcoid lineage and positioned *Indostomus* within the Synbranchiformes.

As nuclear sequences have come to be integrated into molecular analyses, incongruencies between molecular and morphological analyses of gasterosteiform relationships have become more pronounced. Within an extensive acanthomorph data set, Chen *et al.* (2003) sampled the gasterosteoid *Spinachia* and the syngnathoids *Aulostomus* and *Macroramphosus* and analysed sequence data from two nuclear (28*S* rDNA and rhodopsin) and two mitochondrial genes (12*S* and 16*S* rDNA). *Spinachia* was placed basally among a miscellaneous assortment of fishes, while *Aulostomus* and *Macroramphosus* were found to be closely related and situated in a well-supported lineage together with *Dactylopterus*. The analysis of combined mtDNA (12*S* and 16*S* rDNA) and nDNA (28*S* rDNA, histone H3 and TMO-4c4) data sets by Smith & Wheeler (2004) and Smith & Craig (2007) consistently recovered a relationship of gasterosteoids with the cottoid–zoarcoid lineage of Imamura & Yabe (2002). Although *Aulostomus* was the only syngnathoid sampled in these two studies, it was found to be distantly related to gasterosteoids, consistent with the earlier results of Chen *et al.* (2003). Dettai & Lecointre (2005) added taxa and two portions of a novel nuclear locus [mixed lineage leukaemia-like gene (MLL)] to the data set of Chen *et al.* (2003). In all their analyses, the Gasterosteiformes was paraphyletic, and gasterosteoids were placed again at the base of a cottoid–zoarcoid clade distant from syngnathoids. Oddly, while their analysis of the MLL gene recovered a monophyletic Syngnathidae, their total-evidence phylogenetic analysis (including previously published rhodopsin, 28*S* rDNA and mtDNA 12*S* and 16*S* rDNA data) placed *Syngnathus* in a clade distant from all other syngnathoids, including *Hippocampus* (Dettai & Lecointre, 2005), a result that has not been found in any other study of syngnathoid phylogenetics.

Further emphasizing the contrast between molecular and morphological studies of gasterosteiforms, the supertree analysis of Li *et al.*(2009), which sampled four nuclear genes across a broad array of acanthomorphs, placed gasterosteiforms in five different clades: indostomids among synbranchiforms, fistularids together with syngnathids, aulostomids and macroramphosids with dactylopterids, and centriscids in a polytomy among miscellaneous perciforms. In a separate tree, gasterosteoids were placed with the cottoid–zoarcoid lineage. With the exception of gasterosteoid and indostomid relationships, none of these ideas, particularly the paraphyly of syngnathoid superfamilies, are supported by morphological data.

In the most comprehensive molecular study of gasterosteiform relationships to date, Kawahara *et al.* (2008) used complete mitochondrial genome sequences of representatives of all gasterosteiform families (including the Solenostomidae for the first time). Kawahara *et al.*(2008) found the Gasterosteiformes to be paraphyletic, again recovering the Gasterosteoidei together with cottoids and zoarcoids. *Indostomus* was placed with the Synbranchiformes, and members of the Syngnathoidei were placed together with dactylopterids in a polytomy with percomorphs. Kawahara *et al.* (2008) supported a sister group relationship between the Pegasidae and the Syngnathidae and Solenostomidae, as in Orr (1995) [but in contrast with Keivany & Nelson (2006)], while the basal placement of the Centriscoidea by Kawahara *et al.* (2008), and the grouping of the Aulostomoidea with *Dactylopterus* (as the sister of *Aulostomus*) contrasted with both morphological analyses {Orr, 1995; Keivany & Nelson, 2006; though the relative position of these two superfamilies was admittedly only weakly supported in these earlier studies. In contrast to all previous morphological analyses, Kawahara *et al.* (2008) found that the Syngnathoidei was closely related to the Gobioidei.

If the Gasterosteiformes are paraphyletic and *Indostomus* is allied with the Synbranchiformes, what is the sister group of the Syngnathoidei and where is its systematic position? While the Gasterosteoidei appears to be related to the cottoid–zoarcoid lineage, the placement of the Syngnathoidei within the acanthomorph bush remains uncertain. Morphological analyses have provided no alternative to the hypothesis of a sister group relationship between the Gasterosteoidei and the Syngnathoidei, and the results of molecular analyses have been incongruent.

Among molecular analyses, the Syngnathoidei (plus Dactylopteridae) was found to be monophyletic only. Although morphological characters may ultimately be discovered which support a sister group relationship among other acanthomorphs, it is probable that questions related to the origins of the group will require a broadened search for appropriate outgroups (Smith & Craig, 2007). Molecular analyses of deep-level phylogenetic relationships may be susceptible to positively misleading evidence, but the ability to rapidly screen a broad diversity of distantly related taxa means that future studies on gasterosteiform relationships will heavily emphasize this form of data.

Chapter 5

Family-level Phylogenetics

The higher level phylogenetic analyses have proven problematic for syngnathoids, the evolutionary relationships among members of the family Syngnathidae have been resolved with greater confidence. As outlined above, Linnaeus recognised syngnathids in his earliest classification of animals and described both seahorses and pipefishes in the first edition of *Systema Naturae*. Following the descriptions of many new syngnathid species during the 18th and early 19th centuries, the first systematic description of the group was published by Kaup, who considered the location and degree of development of the male brood pouch in dividing the family Syngnathidae into four subfamilies.

In the first explicitly phylogenetic analysis of syngnathid evolution, system of classification to develop an evolutionary hypothesis on the origin and diversification of the group. According to Herald (1959), the Syngnathidae diverged early in its evolution into tail and trunk-brooding species. Following the development of a rudimentary form of male brooding in both these lineages, brood-pouch complexity evolved in parallel in the Gastrophori and Urophori, resulting in the fully enclosed pouch of the seahorse and the highly developed brooding structures found in some gastrophorine species. Herald (1959) suggested that the brooding structures of urophorine pipefish with partially enclosed brood pouches could be further subdivided into monophyletic lineages according to their method of closure (inverted, semi-inverted, overlapping and everted) and proposed a multi-stage model by which the fully enclosed pouch of the seahorse was derived from pipefish ancestors with an everted brooding structure.

The most recent researcher to tackle family-level syngnathid relationships from a global perspective. In revisions of North Atlantic, Indo-Pacific and Mediterranean syngnathids, Dawson made major strides towards clarifying the taxonomy of the group, synonymising many species that had been previously described on the basis of regional collections. Only 294 of the 558 nominal species of syngnathids are presently considered to be valid, and Dawson was responsible for the reclassification of 185 (>30%) of these nominal taxa. Over a career spanning more than five decades, Dawson also described 37 new species of syngnathids, 36 of which are still recognised as valid.

In the first family-level molecular phylogenetic study of the Syngnathidae, Wilson *et al.* (2001) used a suite of three mtDNA markers to reconstruct the evolutionary history of the group. This study included 34 species distributed across 16 of the presently recognised 54 syngnathid genera, with a strong emphasis on the two most species-rich genera in the family, *Hippocampus* and *Syngnathus*. While the taxonomic sampling of this study was far from complete, these first molecular phylogenetic results provided several key insights into the evolution of the family. Wilson *et al.*'s (2001) study supported Duncker's (1912) morphologically based grouping of the Urophori and Gastrophori, indicating that trunk- and tail-brooding lineages diverged early in the evolution of the family, but suggested that several major pouch types within each of these lineages had independent evolutionary origins, challenging the phylogenetic model proposed by Herald (1959). This study also supported a close evolutionary relationship between *Syngnathus* pipefish and *Hippocampus*, contradicting Herald's (1959) theory on the origin of the seahorse brood pouch [though the limited taxonomic sampling of Wilson *et al.* (2001) did not permit a robust test of this hypothesis]. A recently published phylogenetic analysis of syngnathids has provided compelling evidence that trunk brooding may have been secondarily acquired in *Syngnathoides biaculeatus* (Bloch 1785), making this species of particular interest in studies aimed at understanding the functional and morphological changes associated with the evolution of male pregnancy in this group.

Unfortunately, while the mtDNA-based molecular phylogeny of Wilson *et al.* (2001) provides strong support for the early divergence of the Urophori and the Gastrophori and supports the monophyly of several major in-group taxa, much of the backbone of the syngnathid phylogenetic tree remains poorly resolved, complicating efforts to fully understand the pattern of speciation and diversification in the group. Several follow-up studies have broadened the species sampling of this original phylogeny for the three mtDNA genes used in the original Wilson *et al.* (2001) analysis (Wilson *et al.*, 2003; Wilson & Rouse, 2010) and have addressed important questions related to the evolution of the family, but the timing of key evolutionary events during the early evolution of the Urophori remain unresolved. Further resolution of the syngnathid family tree will probably only be achieved with the use of nDNA loci, whose slower rate of evolution may be more appropriate for resolving deeper level divergences in this group. A major phylogenetic analysis of the Syngnathidae using both nDNA and mtDNA markers and including a much wider taxonomic sample of species is currently underway (H. Hamilton, pers. comm.) and is expected to provide higher resolution at the deeper nodes in the syngnathid phylogeny.

Fossil Records

All members of the Syngnathidae have a dermal skeleton composed of body plates, and this group is consequently relatively well represented in the marine fossil record. The availability of a rich fossil record is an essential prerequisite for studies aimed at the evolutionary origins of a group and, together with molecular data, can provide a temporal framework in which to study both its origin and pattern of diversification.

Syngnathoid fossil finds are richest in the Mediterranean Sea and Black Sea regions, but large numbers of Oligocene and Miocene fossils have also been collected from Paratethys sites in central Europe. In addition to this rich European fossil record, fossils of three syngnathid species have been collected from the Modelo and Puente formations of southern California (Fritzsche, 1980). Mirroring the confusion surrounding the higher level relationships among contemporary taxa, considerable debate has surrounded the taxonomic placement of many syngnathoid fossils. Many fossil species remain undescribed, and while the fossil species discussed here are believed to represent valid taxa, a comprehensive consideration of the more problematic gasterosteiform fossils awaits a detailed review.

The oldest syngnathoid fossils date from the early Eocene [48–50 million years before present (B.P.)], and the majority of these fossil taxa have been collected from the rich fossil beds of the Monte Bolca formation in northern Italy (Blot, 1980). Six fossil syngnathoids have been collected from Eocene deposits in this region, representing five distinct taxonomic lineages. Species of the now extinct Ramphosidae, a family which has been suggested to be a sister group to the extant Pegasidae (Pietsch, 1978), have been collected from Eocene sites in Monte Bolca (Blot, 1980) and from a second locality in Jutland, Denmark (Nielsen, 1960). Three additional Monte Bolca fossils are believed to represent morphologically-distinct syngnathoid lineages, and *Solenorhynchus elegans* (Heckel 1853) and *Calamostoma breviculum* Agassiz 1833 are thought to be intermediate forms between the Solenostomidae and the Syngnathidae (Orr, 1995). The oldest syngnathid fossil is believed to be *Prosolenostomus lessenii* Blot 1980, a species that has been placed in the stem lineage of the Syngnathidae on the basis of its absence of dorsal and pelvic fins (Orr, 1995). While the age of *Prosolenostomus* indicates that the family Syngnathidae is at least 50 million years old, the high diversity of syngnathoid fossils found in Eocene deposits suggests that the family probably predates this period.

While the fossils collected from the Eocene deposits of Monte Bolca cannot be confidently assigned to an extant genus, more recent fossils have been partitioned to either the Urophori or the Gastrophori on the basis of their brooding structures. The genus *Hipposyngnathus* is the oldest gastrophorine lineage yet discovered and represents an extinct lineage of pipefishes that persisted from the early Oligocene until the Miocene. Three fossil species have been described from this genus, two from early Oligocene deposits in the Caucasus and Carpathian basins [*Hipposyngnathus convexus* Danil'chenko 1960 and *Hipposyngnathus neriticus*Jerzmañska 1968 (Danil'chenko, 1967; Kotlarczyk *et al.*, 2006)] and one from Miocene deposits of southern California (*Hipposyngnathus imporcitor*Fritzsche 1980), indicating that this genus achieved a wide distribution during its existence.

The earliest gastrophorine fossils with clear affinities for extant genera are the early Oligocene *Dunckerocampus incolumis* Danil'chenko 1967 and *Dunckerocampus squalidus* Danil'chenko 1967 from the Caucasus (Danil'chenko, 1967) and the Miocene *Nerophis zapfei* Bachmeyer 1980 (Bachmayer, 1980) of eastern Austria, indicating the extended evolutionary history of these two lineages. The modern *Nerophis* is found in the temperate coastal habitats along the European coastline; *Dunckerocampus* spp. are currently restricted to the Indo-West Pacific Ocean (Froese & Pauly, 2010).

Both *Hippocampus* and *Syngnathus* are well represented in the urophorine fossil record. *Syngnathus* has the richest fossil record of any extant syngnathid genus, and 11 fossil *Syngnathus* species are currently recognised. The oldest *Syngnathus* fossils date from the early Oligocene (Danil'chenko, 1967; Kotlarczyk *et al.*, 2006), and the genus has an almost uninterrupted fossil record from this period until the present day. The highest diversity of *Syngnathus* fossils has been found in the pre-Messinian deposits of the Miocene Mediterranean Sea, though the existence of *Syngnathus avus* Jordan & Gilbert 1919 in southern Californian formations indicates that the genus has been present on the Pacific Coast of North America for at least 15 million years. While at least four *Syngnathus* species were present in the Miocene Mediterranean Sea (15 M B.P.), subsequent deposits are dominated by a single widespread species (*Syngnathus albyi* Sauvage 1870), which has been collected from many sites around the modern Mediterranean Sea basin (Landini & Sorbini, 2005). Syngnathus *albyi* appears to have persisted through the Messinian salinity crisis in the Mediterranean Sea (*c.* 6 M B.P.), but is replaced in more recent Pliocene and Pleistocene deposits by the modern *Syngnathus acus* L. 1758.

While a Pleistocene fossil of the extant *Hippocampus hippocampus* (L. 1758) was long the only known representative of this genus in the fossil record (Landini & Sorbini, 2005), two Miocene fossils have been described from Sarmatian deposits in Slovenia [*Hippocampus sarmaticus* Zalohar, Hitij & Kriznar 2009 and *Syngnathus slovenicus* Zalohar, Hitij & Kriznar (Zalohar *et al.*, 2009)], considerably extending the age of this lineage. The two Slovenian seahorse fossils are clearly distinct, with *H. sarmaticus* showing affinities for the modern *Hippocampus trimaculatus* Leach 1814 and *H. slovenicus* bearing a resemblance to contemporary pygmy seahorses (Zalohar *et al.*, 2009). Again, major morphological differences between these two fossils suggest that the seahorse lineage predates the Miocene, consistent with recent molecular clock reconstructions of the group.

Intrageneric Relationships

Both phylogenetic and phylogeographic studies have heavily emphasized the evolution of seahorses (genus *Hippocampus*) and *Syngnathus* pipefishes, providing insights into the biogeography of these two genera. A 2004 issue of *Molecular Phylogenetics and Evolution* saw the publication of back-to-back articles on the evolutionary relationships of the seahorse. The first, by Casey *et al.* (2004), analysed full-length cytochrome *b* sequences in a sample of 22 seahorse species, including multiple representatives of each species sampled. A second study by Teske *et al.* (2004) supplemented the Casey *et al.* (2004) cytochrome *b* data set with 16*S* rDNA sequence data and two nuclear genes (*RP*1 and *aldolase*) and included a total of 30 seahorse species. The analysis of Teske *et al.* (2004) produced a highly resolved phylogeny of the seahorse genus and provided important insights into the evolutionary history of the group.

On the basis of the placement of the Australian *Hippocampus breviceps* Peters 1869 and Australian and New Zealand *Hippocampus abdominalis* Lesson 1827 at the base of the seahorse phylogeny and the south-west Pacific Ocean distribution of several outgroup species, Teske *et al.* (2004) hypothesized an Australian origin for the genus *Hippocampus*. The main group of seahorses falls into three major clades, two with Indo-Pacific affinities and one with a circumglobal distribution, a lineage which is thought to have had its origin following the colonisation of the Atlantic Ocean (Teske *et al.*, 2004). This lineage subsequently recolonised the Indo-West Pacific Ocean, giving rise to a species-rich group including the widespread *Hippocampus kuda* Bleeker 1852. The alternative hypothesis, that Atlantic Ocean seahorse species are the descendents of repeated colonisations of this region, however, could not be statistically rejected. Subsequent biogeographic analyses indicate that long-distance dispersal events have placed an important role in the origin of new seahorse lineages in the Indo-West Pacific (Teske *et al.*, 2005) and Atlantic Oceans.

The original family-level phylogeny of the Syngnathidae included 11 species of *Syngnathus* pipefishes from Europe, Asia and North America. While the relationships among *Syngnathus* pipefishes were poorly resolved using the pruned data set published in the original paper (third codons of cytochrome *b* were removed due to mutational saturation at high genetic distances), the full data set used for the updated family-level phylogeny published here provides a high degree of resolution within *Syngnathus*, with strong support for all major nodes in the *Syngnathus* phylogeny.

Similar to the results observed for *Hippocampus*, there is clear geographical structure in *Syngnathus* biogeography. The most basal lineages in the *Syngnathus* phylogeny are restricted to the Pacific coast of North America [*Syngnathus exilis* (Osburn & Nichols 1916) and*Syngnathus leptorhynchus* Girard 1854], and the remainder of the phylogeny is divided into three major clades, a widespread Atlantic coast group [*Syngnathus fuscus* Storer 1839 and *Syngnathus scovelli* (Everman & Kendall 1896)], a second clade containing Atlantic coast and Caribbean species, and a third clade containing the western Pacific *Syngnathus schlegeli* Kaup 1856 and a monophyletic clade of European pipefishes. The close phylogenetic relationships between clades on east and west coasts of North America and Eurasia suggest that interoceanic dispersal may be relatively common in *Syngnathus*, a pattern which mirrors that detected in *Hippocampus*. Interestingly, given that the oldest *Syngnathus* fossils are from Oligocene sites in eastern Europe, the present-day phylogeny of the genus appears to reflect the consequences of extensive speciation and extinction during the evolution of the group.

Phylogeography

Recent publications concerning the evolution and origin of syngnathids have been dominated by phylogeographic investigations of individual species. Here again, studies have focused almost exclusively on *Hippocampus* and *Syngnathus*, providing an excellent data set in which to identify general patterns in the contemporary phylogeography of these two genera.

Despite differences in the spatial scales (from 10^1 to 10^4 km), the number of individuals screened (from <10 to >300) and the markers used (mtDNA and nDNA sequence data, microsatellites and allozymes), the majority of studies on both seahorses and pipefishes have found moderate-to-high genetic diversity and strong phylogeographic structure, consistent with large effective population sizes and low dispersal potential of syngnathid species. Studies conducted on northern-hemisphere temperate water species have typically found evidence of recent population expansions, consistent with the colonisation of northern habitats following the end of the last glacial maximum (*c.* 20 k B.P.). The large number of phylogeographic studies published over the past decade cannot be reviewed in detail here, but a number of case studies serve to illustrate how phylogeographic analyses have contributed to a greater understanding of syngnathid evolution.

Panmixia in Populations of *H. Abdominalis*

In contrast to the majority of phylogeographic studies on syngnathid species, which have found evidence of strong population structuring in both seahorses and pipefishes, recent investigations by Armstrong (2001) and Nickel (2009) found a surprising lack of phylogeographic structure in New Zealand and Australian populations of *H. abdominalis*. Armstrong (2001) used a combination of morphological characters, mtDNA sequence data and allozymes to investigate the hypothesis that Australian and New Zealand populations of *H. abdominalis* represent distinct species (Kuiter, 2001). While morphological variation in *H. abdominalis* was high, the majority of this variation was found within populations, with no diagnostic differences between the putative species. Allozyme markers could not distinguish between New Zealand and Australian populations, though some evidence of geographic structure was detected in mtDNA data, where the majority of New Zealand individuals formed a monophyletic clade. Armstrong (2001) rejected the hypothesis that *H. abdominalis* represented more than a single species, though molecular data do appear to indicate that populations in these two regions may be in the early stages of divergence.

In a study aimed at investigating the conservation status of New Zealand populations of *H. abdominalis*, Nickel (2009) used multiple mtDNA markers and nuclear microsatellites to investigate the phylogeographic structure of a large number of populations on both the north and south islands of the country. Although Nickel (2009) observed a remarkable range of morphological and colour variation among individuals [consistent with the results of Armstrong (2001)], and high genetic diversity at both mtDNA and microsatellite loci, no evidence of phylogeographic structure was found. The distribution of mtDNA haplotypes was found to be consistent with a recent population expansion, leading Nickel (2009) to suggest that New Zealand may have been colonised relatively recently by Australian seahorses, a situation which could help to explain the lack of population structure observed in the region. The application of molecular dating to these data should permit a robust test of timing of the colonisation of New Zealand by *H. abdominalis*.

Identifying Phylogeographic Breaks: The Importance of Scale

The broad distributions of many syngnathids make this group particularly well suited for studying the population genetic consequences of environmental variation across time and space. The Pacific Coast pipefish *S. leptorhynchus* is one of the most widespread pipefish species and is distributed across almost 4000 km of coastline between the Alexander Archipelago, Alaska, U.S.A. and Baja California, Mexico. Morphological variation in *S. leptorhynchus* is high, a finding which has led some researchers to suggest that it may actually comprise more than a single species (Herald, 1941). On the basis of meristic variation, Herald (1941) suggested that *S. leptorhynchus*could be divided into a northern and southern species, with a biogeographic boundary near Point Conception in southern California.

Two recent studies have used molecular markers to study the Pacific Coast phylogeography of *S. leptorhynchus* (Louie, 2003; Wilson, 2006). While both studies covered the same geographic range, Louie (2003) used a high-density sampling strategy, surveying almost 400 individuals sampled from 29 coastal sites, while Wilson (2006) sampled larger numbers of individuals from four collection localities. Both studies found evidence of the post-glacial colonisation of northern sites following the end of the last glacial cycle and detected evidence of a phylogeographic break between sites in the north-west Pacific

Ocean and those farther south. Interestingly, while the phylogeographic break identified by Wilson (2006) was found between central Oregon and California, U.S.A., the mtDNA data of Louie (2003) suggested that this break was located nearly 500 km farther north, in northern Washington. While the statistical approaches used in these two studies were not identical, differences in their conclusions can be largely attributed to differences in their sampling strategies, illustrating the potential challenges inherent in interpreting the results of molecular analyses.

Comparative Phylogeography of Seahorses: New insights Via Species-level Comparisons

In what is undoubtedly the most impressive analysis of comparative phylogeographic structure in a group of syngnathids, Lourie *et al.*(2005) investigated the phylogeographic structure of four south-east Asian seahorse species inhabiting a region dominated by the presence of the Sunda Shelf, a large shallow sea which has been repeatedly exposed as land during the sea-level reductions associated with Pleistocene glacial cycles. Lourie *et al.* (2005) sampled widely (11–29 populations of each species) and, most importantly, used the same fragment of the mtDNA cytochrome *b* gene for all species, facilitating the direct comparison of phylogeographic structure. Differences in the contemporary population genetic structure of the four *Hippocampus* species could be directly attributed to species-specific habitat and ecological differences. While genetic diversity was high in all species, population genetic structuring was especially pronounced in *Hippocampus barbouri* Jordan & Richardson 1905 and *H. kuda*, two shallow-water species, while phylogeographic structure was more modest in the deep-water *Hippocampus spinosissimus* Weber 1913 and *H. trimaculatus*.

Differences in phylogeographic structure among these species appear to reflect differences in their relative dispersal abilities. While both deep-water species appear to have colonised the Sunda Shelf since its most recent flooding (*c.* 15 k B.P.), *H. barbouri* has not yet spread into the region, and the large number of private haplotypes found in shelf populations of *H. kuda* suggest that these lineages are derived from a refugial population. In an innovative use of haplotype frequency data, Lourie *et al.* (2005) inferred dispersal potential of the four study species by examining the geographic distribution of individual haplotypes, an approach which serves as an indirect estimate of relative dispersal. The scale of the Lourie *et al.* (2005) investigation exceeds that of most previously published phylogeographic studies on the group [Woodall (2009) provides a comparably impressive data set for European seahorses], and such comparative studies offer one of the best means by which the factors contributing to population and species-level divergence can be identified.

Researchers have long been interested in the evolutionary relationships among syngnathids, and the past several decades have provided new insights into the evolution of the group. The application of molecular methods has proven particularly effective in illuminating more recent evolutionary events, while both molecular and morphological analyses have failed to provide a well-resolved model of deeper level relationships, complicating efforts to understand the evolutionary origins of the family Syngnathidae. Although the sister group relationship of the Solenostomidae and the Syngnathidae appears to be robust, the relationships among other major groups within the order Gasterosteiformes remain poorly understood. Comparative ontogenetic studies may help to complement morphological analyses based on adult specimens, and a broadened analysis using molecular methods may provide an evolutionary hypothesis that could be corroborated with further morphological investigations.

While the large number of recent studies on syngnathid phylogeography and phylogenetics (and on many other aspects of syngnathid biology) have clarified several aspects of syngnathid evolution, most of these studies have been restricted to the two most species-rich genera of the family, seahorses (genus *Hippocampus*), and pipefishes of the genus *Syngnathus*. Although these publications have been prompted in part by considerable conservation interest in seahorse species (Foster & Vincent, 2004), they tend to overlook the majority of morphological and ecological variation found in the rest of the family. As many syngnathid species inhabit the same sensitive environments inhabited by seahorses, they offer potentially useful comparative systems in which to investigate how ecological variation influences species susceptibility to habitat degradation. At the same time, a wider focus on the diversity of forms found in this family promises to provide a deeper understanding of the evolutionary pressures which have influenced its diversification.

Swimming

Aquatic Locomotion

Swimming is biologically propelled motion through a liquid medium. Swimming has evolved a number of times in a range of organisms ranging from arthropods to fish to molluscs.

Evolution of Swimming

Swimming evolved a number of times in unrelated lineages, and the evolutionary pressures leading to its adoption are unknown. Supposed jellyfish fossils occur in the Ediacaran, but the first free-swimming animals appear in the Early to Middle Cambrian. These are mostly related to the arthropods, and include the Anomalocaridids, which swam by means of lateral lobes in a fashion reminiscent of today's cuttlefish. Cephalopods joined the ranks of the nekton in the late Cambrian, and chordates were probably swimming from the Early Cambrian. Many terrestrial animals retain some capacity to swim, however some have returned to the water and developed the capacities for aquatic locomotion.

Invertebrates

Among the radiata, jellyfish and their kin, the main form of swimming is to flex their cup shaped bodies. All jellyfish are free-swimming, although many of these spend most of their time swimming passively. Passive swimming is akin to gliding; the organism floats, using currents where it can, and does not exert any energy into controlling its position or motion. Active swimming, in contrast, involves the expenditure of energy to travel to a desired location.

In bilateria, there are many methods of swimming. The arrow worms (chaetognatha) undulate their finned bodies, not unlike fish. Nematodes swim by undulating their finless bodies. Some Arthropod groups can swim. Many crustaceans can swim. Most, such as shrimp, will usually swim by paddling with special swimming legs (pleopods), however some arthropods can also propel themselves backwards quickly by flicking their tail, known as the Caridoid escape reaction. Swimming crabs swim with modified walking legs (pereiopods). Daphnia, an crustacean, swims by beating it antenna instead. There are also a number of forms of swimming molluscs. Many free-swimming sea slugs, such as sea angels, flap finlike structures. Some shelled molluscs, such as scallops can briefly swim by clapping their two shells open and closed. The molluscs most evolved for swimming are the cephalopods.

All cephalopods can move by jet propulsion, but this is a very energy-consuming way to travel compared to the tail propulsion used by fish. The relative efficiency of jet propulsion decreases further as animal size increases. Since the Paleozoic, as competition with fish produced an environment where efficient motion was crucial to survival, jet propulsion has taken a back role, with fins and tentacles used to maintain a steady velocity. The stop-start motion provided by the jets, however, continues to be useful for providing bursts of high speed - not least when capturing prey or avoiding predators. Indeed, it makes cephalopods the fastest marine invertebrates, and they can out accelerate most fish. Oxygenated water is taken into the mantle cavity to the gills and through muscular contraction of this cavity, the spent water is expelled through the hyponome, created by a fold in the mantle. Motion of the cephalopods is usually backward as water is forced out anteriorly through the hyponome, but direction can be controlled somewhat by pointing it in different directions. Most cephalopods float (i.e. are neutrally buoyant), so do not need to swim to remain afloat.

Jet propulsion

Jet propulsion is a method of aquatic locomotion where animals fill a muscular cavity and squirt out water to propel them in the opposite direction of the squirting water. Most organisms are equipped with one of two designs for jet propulsion; they can draw water from the rear and expel it from the rear, such as jellyfish, or draw water from front and expel it from the rear, such as salps. Filling up the cavity causes an increase in both the mass and drag of the animal. Because of the expanse of the contracting cavity, the animal's velocity fluctuates as it moves through the water, accelerating while expelling water and decelerating while vacuuming water. Even though these fluctuations in drag and mass can be ignored if the frequency of the jet-propulsion cycles is high enough, jet-propulsion is a relatively inefficient method of aquatic locomotion.

Hydrozoan medusae, which use a one-way water cavity design, generate a phase of continuous cycles of jet-propulsion followed by a phase of rest. The Froude efficiency is measured to be at 0.09, which shows a very costly method of locomotion. The metabolic cost of transport for the medusa is high when compared to a fish of equal mass. Other jet-propelled animals have similar problems in efficiency. Scallops, which use a similar design, swim by quickly opening and closing their shells, which draws in water and expels it from all sides. This locomotion is used as a means to escape predators such as starfish. Afterwards, the shell acts as a hydrofoil to counteract the scallop's tendency to sink. The Froude efficiency is low for this type of movement, about 0.3, hence why it's used as an emergency escape mechanism from predators. However, the amount of work the scallop has to do is mitigated by the elastic hinge that connects the two shells of the bivalve. Squids swim by drawing and expelling water through their siphon and into their mantle cavity. The Froude efficiency of their jet-propulsion system is around 0.29, which is much lower than a fish of the same mass. Squid swim more slowly than fish, but use more power to generate their speed. The loss in efficiency is due to the amount of water the squid can accelerate out of its mantle cavity.

Elasticity in Jet Propulsion

Much of the work done by Scallop muscles to close its shell is stored as elastic energy in abductin tissue, which acts as a spring to open the shell. The elasticity causes the work done against the water to be low because of the large openings the water has to enter and the small openings the water has to leave. The inertial work of scallop jet-propulsion is also low. Because of the low inertial work, the energy savings created by the elastic tissue is so small that it's negligible. Medusae can also use their elastic mesoglea to enlarge their bell. Their mantle contains a layer of muscle sandwiched between elastic fibres. The muscle fibres run around the bell circumferentially while the elastic fibres run through the muscle and along the sides of the bell to prevent lengthening. After making a single contraction, the bell vibrates passively at the resonant frequency to refill the bell. However, in contrast with scallops, the inertial work is similar to the hydrodynamic work due to how medusas expel water - through a large opening at low velocity. Because of this, the negative pressure created by the vibrating cavity is lower than the positive pressure of the jet, meaning that inertial work of the mantle is small. Thus, jet-propulsion is shown as an inefficient swimming technique.

Swimming in Fish

Many fish swim through water by creating undulations with their bodies or oscillating their fins. The undulations create components of forward thrust complemented by a rearward force, side forces which are wasted portions of energy, and a normal force that is between the forward thrust and side force. Different fish swim by undulating different parts of their bodies. Eel-shaped fish undulate their entire body in rhythmic sequences. Streamlined fish, such as salmon, undulate the caudal portions of their bodies. Some fish, such as sharks, use stiff, strong fins to create dynamic lift and propel themselves. It is common for fish to use more than one form of propulsion, although they will display one dominant mode of swimming Gait changes have even been observed in juvenile reef fish of various sizes. Depending on their needs, fish can rapidly alternate between synchronised fin beats and alternating fin beats

Body-Caudal Fin (BCF) Propulsion

- Aguilliform: Anguilliform swimmers are typically slow swimmers. They undulate the majority of their body and use their head as the fulcrum for the load they are moving. At any point during their undulation, their body has an amplitude between 0.5-1.0 wavelengths. The amplitude that they move their body through allows them to swim backwards. Anguilliform locomotion is usually seen in fish with long, slender bodies like eels, lampreys, many varieties of larvae, and oar fish.
- Subcarangiform, Carangiform, Thunniform: These swimmers undulate the posterior half of their body and are much faster than anguilliform swimmers. At any point while they are swimming, a wavelength <1 can be seen in the undulation pattern of the body. Some Carangiform swimmers include nurse sharks, bamboo sharks, and reef sharks. Thunniform swimmers are very fast and some common Thunniform swimmers include tuna, white sharks, salmon, jacks, and mako sharks. Thunniform swimmers only undulate their high aspect ratio caudal fin, so they are usually very stiff to push more water out of the way.
- Ostraciform: Ostraciform swimmers oscillate their caudal region, making them relatively slow swimmers. Boxfish, torpedo rays, and momyrs employ Ostraciform locomotion. The cow fish uses Osctraciform locomotion to hover in the water column.

Median Paired Fin (MPF) Propulsion

- Tetraodoniform, Balistiform, Diodontiform: These swimmers oscillate their median fins. They are typically slow swimmers, and some notable examples include the oceanic sunfish (which has extremely modified anal and dorsal fins), puffer fish, and triggerfish.
- Rajiform, Amiiform, Gymnotiform: This locomotory mode is accomplished by undulation of the pectoral and median fins. During their undulation pattern, a wavelength >1 can be seen in their fins. They are typically slow to moderate swimmers, and some examples include rays, bowfin, and knife fishes. The black ghost knife fish is a Gymnotiform swimmer that has a very long ventral ribbon fin. Thrust is produced by passing

waves down the ribbon fin while the body remains rigid. This also allows the ghost knife fish to swim in reverse.

- Labriform: Labriform swimmers are also slow swimmers. They oscillate their pectoral fins to create thrust. Oscillating fins create thrust when a starting vortex is shed from the trailing edge of the fin. As the foil departs from the starting vortex, the effect of that vortex diminishes, while the bound circulation remains, producing lift. Labriform swimming can be viewed as continuously starting and stopping. Wrasses and surf perch are common Labriform swimmers

Hydrofoils

Hydrofoils, or fins, are used to push against the water to create a normal force to provide thrust, propelling the animal through water. Sea turtles and penguins beat their paired hydrofoils to create lift. Some paired fins, such as pectoral fins on leopard sharks, can be angled at varying degrees to allow the animal to rise, fall, or maintain its level in the water column. The reduction of fin surface area helps to minimise drag, and therefore increase efficiency. Regardless of size of the animal, at any particular speed, maximum possible lift is proportional to (wing area) x $(\text{speed})^2$. Dolphins and whales have large, horizontal caudal hydrofoils, while many fish and sharks have vertical caudal hydrofoils. Porpoising (seen in cetaceans, penguins, and pinnipeds) may save energy if they are moving fast. Since drag increases with speed, the work required to swim unit distance is greater at higher speeds, but the work needed to jump unit distance is independent of speed. Seals propel themselves through the water with their caudal tail, while sea lions create thrust solely with their pectoral flippers.

Drag Powered Swimming

As with moving through any fluid, friction is created when molecules of the fluid collide with organism. The collision causes drag against moving fish, which is why many fish are streamlined in shape. Streamlined shapes work to reduce drag by orienting elongated objects parallel to the force of drag, therefore allowing the current to pass over and taper off the end of the fish. This streamlined shape allows for more efficient use of energy locomotion. Some flat-shaped fish can take advantage of pressure drag by having a flat bottom surface and curved top surface. The pressure drag created allows for the upward lift of the fish.

Appendages of aquatic organisms propel them in two main and biomechanically extreme mechanisms. Some use lift powered swimming, which can be compared to flying as appendages flap like wings, and reduce drag on the surface of the appendage. Others use drag powered swimming, which can be compared to oars rowing a boat, with movement in a horizontal plane, or paddling, with movement in the parasagittal plane.

Drag swimmers use a cyclic motion where they push water back in a power stroke, and return their limb forward in the return or recovery stroke. When they push water directly backwards, this moves their body forward, but as they return their limbs to the starting position, they push water forward, which will thus pull them back to some degree, and so opposes the direction that the body is heading. This opposing force is called drag. The return-stroke drag causes drag swimmers to employ different strategies than lift swimmers. Reducing drag on the return stroke is essential for optimising efficiency. For example, ducks paddle through the water spreading the webs of their feet as they move water back, and then when they return their feet to the front they pull their webs together to reduce the subsequent pull of water forward. The legs of water beetles have little hairs which spread out to catch up and move water back in the power stroke, but lay flat as the appendage moves forward in the return stroke. Also, the water beetle's legs have a side that is wider and is held perpendicular to the motion when pushing backward, but the leg is then rotated when the limb is to return forward, so that the thinner side will catch up less water.

Drag swimmers experience a lessened efficiency in swimming due to resistance which affects their optimum speed. The less drag a fish experiences, the more it will be able to maintain higher speeds. Morphology of the fish can be designed to reduce drag, such as streamlining the body. The cost of transport is much higher for the drag swimmer, and when deviating from its optimum speed, the drag swimmer is energetically strained much more than the lift swimmer. There are natural processes in place to optimise energy use, and it is thought that adjustments of metabolic rates can compensate in part for mechanical disadvantages.

Semi-aquatic animals compared to fully aquatic animals exhibit exacerbation of drag. Design that allows them to function out of the water limits the efficiency possible to be reached when in the water. In water swimming at the surface exposes them to resistive wave drag and is associated with a higher cost than submerged swimming.

Swimming below the surface exposes them to resistance due to return strokes and pressure, but primarily friction. Frictional drag is due to fluid viscosity and morphology characteristics. Pressure drag is due to the difference of water flow around the body and is also affected by body morphology. Semi-aquatic organisms encounter increased resistive forces when in or out of the water, as they are not specialised for either habitat. The morphology of otters and beavers, for example, must meet needs for both environments. Their fur decreases streamlining and creates additional drag. The platypus may be a good example of an intermediate between drag and lift swimmers because it has been shown to have a rowing mechanism which is similar to lift-based pectoral oscillation. The limbs of semi-aquatic organisms are reserved for use on land and using them in water not only increases the cost of locomotion, but limits them to drag-based modes. Although they are less efficient, at low speeds drag swimmers are able to produce more thrust than lift swimmers. They are also thought to be better for maneuverability due to the large thrust produced.

Fast Starts and Mauthner Response

Varieties of fish, such as teleosts, use fast-starts to escape from predators. Fast-starts are characterised by the muscle contraction on one side of the fish twisting the fish into a C-shape. Afterwards, muscle contraction occurs on the opposite side to allow the fish to enter into a steady swimming state with waves of undulation travelling alongside the body. The power of the bending motion comes from fast-twitch muscle fibres located in the central region of the fish. The signal to perform this contraction comes from a set of Mauthner cells which simultaneously send a signal to the muscles on one side of the fish. Mauthner cells are activated when something startles the fish and can be activated by visual or sound-based stimuli.

Fast-starts are split up into three stages. Stage one, which is called the preparatory stroke, is characterised by the initial bending to a C-shape with small delay caused by hydrodynamic resistance. Stage two, the propulsive stroke, involves the body bending rapidly to the other side, which may occur multiple times. Stage three, the rest phase, cause the fish to return to normal steady-state swimming and the body undulations begin to cease. Large muscles located closer to the central portion of the fish are stronger and generate more force than the muscles in the tail. This asymmetry in muscle composition causes body undulations that occur in Stage 3. Once the fast-start is completed, the position of the fish has been shown to have a certain level of unpredictability, which helps fish survive against predators.

The rate at which the body can bend is limited by resistance contained in the inertia of each body part. However, this inertia assists the fish in creating propulsion as a result of the momentum created against the water. The forward propulsion created from C-starts, and steady-state swimming in general, is a result of the body of the fish pushing against the water. Waves of undulation create rearward momentum against the water providing the forward thrust required to push the fish forward.

Efficiency

Froude propulsion efficiency: ratio of power output to input

$$\eta f = 2U1(U1 + U2)$$

U1 = free stream velocity, U2 = jet velocity

A good efficiency for Carangiform propulsion is between 50-80%

Minimising Drag

Pressure differences occur outside the boundary layer of swimming organisms due to disrupted flow around the body. The difference on the up- and down-stream surfaces of the body is pressure drag, which creates a downstream force on the object. Frictional drag, on the other hand, is a result of fluid viscosity in the boundary layer. Higher turbulence causes greater frictional drag.

Reynolds number (Re) is the measure of the relationships between inertial and viscous forces in flow ((animal's length x animal's velocity)/ kinematic viscosity of the fluid). Turbulent flow can be found at higher Re values, where the boundary layer separates and creates a wake, and laminar flow can be found at lower Re values, when the boundary layer separation is delayed, reducing wake and kinetic energy loss to opposing water momentum.

The body shape of a swimming organism affects the resulting drag. Long, slender bodies reduce pressure drag by streamlining, while short, round bodies reduce frictional drag; therefore, the optimal shape of an organism depends on its niche. Swimming organisms with a fusiform shape are likely to experience the greatest reduction in both pressure and frictional drag.

Wing shape also affects the amount of drag experienced by an organism, as with different methods of stroke, recovery of the pre-stroke position results in the accumulation of drag.

High-speed ram ventilation creates laminar flow of water from the gills along the body of an organism.

The secretion of mucus along the organism's body surface, or the addition of long-chained polymers to the velocity gradient, can reduce frictional drag experienced by the organism.

Buoyancy

Many aquatic/marine organisms have developed organs to compensate for their weight and control their buoyancy in the water. These structures, make the density of their bodies very close to that of the surrounding water. Some hydrozoans, such as siphonophores, has gas-filled floats; the Nautilus, Sepia, and Spirula (Cephalopods) have chambers of gas within their shells; and most teleost fish and many lantern fish (Myctophidae) are equipped with swim bladders. Many aquatic and marine organisms may also be composed of low-density materials. Deep-water teleosts, which do not have a swim bladder, have few lipids and proteins, deeply ossified bones, and watery tissues that maintain their buoyancy. Some sharks' livers are composed of low-density lipids, such as hydrocarbon squalene or wax esters (also found in Myctophidae without swim bladders), which provide buoyancy.

Swimming animals that are denser than water must generate lift or adapt a benthic lifestyle. Movement of the fish to generate hydrodynamic lift is necessary to prevent sinking. Often, their bodies act as hydrofoils, a task that is more effective in flat-bodied fish. At a small tilt angle, the lift is greater for flat fish than it is for fish with narrow bodies. Narrow-bodied fish use their fins as hydrofoils while their bodies remain horizontal. In sharks, the heterocercal tail shape drives water downward, creating a counteracting upward force while thrusting the shark forward. The lift generated is assisted by the pectoral fins and upward-angle body positioning. It is supposed that tunas primarily use their pectoral fins for lift.

Buoyancy maintenance is metabolically expensive. Growing and sustaining a buoyancy organ, adjusting the composition of biological makeup, and exerting physical strain to stay in motion demands large amounts of energy. It is proposed that lift may be physically generated at a lower energy cost by swimming upward and gliding downward, in a "climb and glide" motion, rather than constant swimming on a plane.

Secondary Evolution of Swimming

While tetrapods lost many of their natural adaptations to swimming when they evolved onto the land, many have re-evolved the ability to swim or have indeed returned to a completely aquatic lifestyle. Primarily or exclusively aquatic animals have re-evolved from terrestrial tetrapods multiple times: examples include amphibians such as newts, reptiles such as crocodiles, sea turtles, ichthyosaurs, plesiosaurs and mosasaurs, marine mammals such as whales, seals and otters, and birds such as penguins. Many species of snakes are also aquatic and live their entire lives in the water. Even though primarily terrestrial tetrapods have lost many of their adaptations to swimming, the ability to swim has been preserved or redeveloped in many of them. It may never have been completely lost.

Examples are: Some breeds of dog swim recreationally. Umbra, a world record-holding dog, can swim 4 miles (6.4 km) in 73 minutes, placing her in the top 25% in human long-distance swimming competitions. Although most cats hate water, adult cats are good swimmers. The fishing cat is one wild species of cat that has evolved special adaptations for an aquatic or semi-aquatic lifestyle webbed digits. Tigers and some individual jaguars are the only big cats known to go into water readily, though other big cats, including lions, have been observed swimming. A few domestic cat breeds also like swimming, such as the Turkish Van. In an unpublished research carried out 2002 at the University of Bern (Switzerland), Bender & Hirt showed that the Turkish Van has less inhibition to enter in shallow water compared to another breed, the Russian Blue. This behaviour can be partially explained by the character of the Turkish Van, who seems to be more curious and enterprising than other cat breeds. Horses, moose, and elk are very powerful swimmers, and can travel long distances in the water. Elephants are also capable of swimming, even in deep waters. Eyewitnesses have confirmed that camels, including Dromedary and Bactrian camels, can swim, despite the fact that there is little deep water in their natural habitats. Both domestic and wild rabbits can swim. Domestic rabbits are sometimes trained to swim as a circus attraction. A wild rabbit famously swam in an apparent attack on U.S. President Jimmy Carter's boat when it was threatened in its natural habitat. The Guinea pig (or cavy) is noted as having an excellent swimming ability. Mice can swim quite well. They do panic when placed in water, but many lab mice are used in the Morris water maze, a test to measure learning. When mice swim, they use their tails like flagella and kick with their legs. Many snakes are excellent swimmers as well. Large adult anacondas spend the majority of their time in the water, and have difficulty moving on land.

Humans do not swim instinctively, but nonetheless often feel attracted to water, showing a broader range of swimming movements than other non-aquatic animals. In contrast, many monkeys can naturally swim and some, like the proboscis monkey, crab-eating macaque, and Rhesus macaque swim regularly. Large primates other than humans generally do not like to swim. Wild chimpanzees and some gorillas will wade in very shallow water but will make no attempt to cross larger bodies of water. Orangutans don't swim instinctively but will attempt it under pressure or if learned. Among invertebrates, a number of insect species have adaptations for aquatic life and locomotion. Examples of aquatic insects include dragonfly larvae, water boatmen, and diving beetles. There are also aquatic spiders, although they tend to prefer other modes of locomotion under water than swimming proper.

Human Swimming

Swimming has been known amongst humans since prehistoric times; the earliest record of swimming dates back to Stone Age paintings from around 7,000 years ago. Competitive swimming started in Europe around 1800 and was part of the first modern 1896 Summer Olympics in Athens, though not in a form comparable to the contemporary events. It was not until 1908 that regulations were implemented by the International Swimming Federation to produce competitive swimming.

Chapter 6

Sound Producing Organs

Electrocommunication

Electrocommunication is the communication method used by weakly electric fishes. Weakly electric fishes are a group of animals that utilises a communicating channel that is "invisible" to most other animals: electric signalling. Electric fishes communicate in a process called electrocommunication by generating electric fields and interpreting the signal frequencies, waveforms, and delay, etc. The best studied species that conduct electrocommunication are two freshwater lineages- the African Mormyrids and and the South American Gymnotiformes. While weakly electric fish is the only group that has been identified to carry out both generation and reception of electric fields, other species were also identified to perform either ability (but not both). Animals that either generate or receive electric fields are found only in aquatic (or at least moist) environments due to large resistance of all other media (ex. air) So far, communication between electric fish has been identified mainly to serve the purpose of conveying information on

- species recognition
- courtship and sex differentiation
- motivational status (warning and territory protection) and
- environmental resources.

Overview of Weakly Electric Fish

Electric fish is a general term used to describe fish that can generate external electric fields or receive electric fields (electroreception). Electric fish can be further divided into three categories: strongly discharging, weakly discharging, and fish that can only sense but are unable to generate electric fields Strongly electric fish generates strong electric field up to 500 volts for predatory purposes; strongly electric fish includes both marine and fresh water fishes (two freshwater taxa- African electric catfish (*Malapterurus electricus*) and the Neotropical electric eel (*Electrophorus electricus*) and the marine torpedo rays (*Torpedo*)). Weakly electric fish generates electric fields mainly for communicational and electrolocational purposes; weakly electric fish is found in fresh water only and include African freshwater Mormyridae and Gymnarchus and Neotropical electric knifefishes Lastly, fish that are only able to detect electrical signals includes sharks, rays, and skates. Electric fishes generate discharge from electric organs located near the tail region of the fish. Electric organs are mostly derived from muscle cells (myogenic) with one exception of organs derived from neurons (neurogenic organs) in one gymnotiform family. To detect the electric signals, electric fish has two types of receptive cells- ampullary and tuberous electroreceptors.

Receptive Organs

All organisms respond to sufficiently strong electric shocks, but only some aquatic vertebrates can detect and utilise weak electric field; these organisms are therefore called electroreceptive (For an example, we human-beings react to strong electric currents with a sense of pain and sometimes a mixture of other senses; however, we cannot detect weakly electric fields and therefore are not electroreceptive.) The ability to sense and utilise electric fields was found almost solely in lower, aquatic vertebrates (fishes and some amphibians). Terrestrial animals, with very few exceptions, lack this electric sensing channel due to low conductivity of air, soil, or media other than aqueous environment. Exception in terrestrial vertebrates include Australia monotremes and platypus that sense electric fields generated by fish for hunting.

In order to detect weakly electric fields, animals must possess electroreceptors, or receptive organs, that detect potential differences. For electric fishes, receptive organs are groups of sensory cells rooted in epidermal pits that look like small spots on the skin. In each receptive organ, there are sensory cells embedded in the bottom of the opened "pit" that faces outside. Electroreceptors detect electric signals by building up a potential difference between the outside environment and the fish body's internal environment; current flow due to such potential difference further results in a receptor potential that's presynaptic to the sensory fibres. Finally, this receptor potential leads to action potential fired by sensory cells.

Electric fishes carry a variety of sensitive receptive organs that are tuned to different types and ranges of signals. To classify types of electroreceptors, the first branch should be made between ampullary and tuberous organs, which exist in both mormyrids and gymnotiforms. These two types of electric receptors have very distinct anatomical differences- ampullary organs have their opened "pit" formed in a duct-like structure and filled with mucous substance; meanwhile, the "pit" of a tuberous organ is loosely packed with epithelial cells. In addition to anatomical differences, these two receptors also have distinctive functional differences. Ampullary organs are more sensitive and tuned to a *low frequency range* of 1-10Hz, which is the range of non-electrogenic, biological source of electricity; therefore, ampullary organs are mainly used for passive electrolocation. On the other hand, tuberous, which are used for electrocommunication by weakly electric fishes, are less sensitive and tuned to much *higher frequencies.*

Classification of the Two Types of Receptive Organs

Type	Structure	Function	Sensitivity	Where found
Ampullary	Open pit/ Filled with mucous	Electrolocation/ Locate preys	High: 0.01 μV/cm in marine species, 0.01 mV/cm in freshwater; Sensitive to DC fields/ Low frequencies less than 50 Hz	Sharks & Rays; Non-teleost fishes; Certain teleosts (mormyrids, certain notopterus, gymnotiforms, catfish); Amphibians (except frogs and toads)
Tuberous	Covered by skin - loosely packed with epithelia cells	Electrocommunication	0.1 mV to 10 mV/cm/ Tens of Hz to more than 1 kHz.	Mormyrid fish; Gymnotiform fish

Tuberous organs

Tuberous organs, the type of receptive organ used for electrocommunication, can be divided into two types depending on the way information is encoded: time coders and amplitude coders. There are multiple forms of each type of tuberous organs, and all weakly electric fish species possess at least one form of both time coders and amplitude coders. Time coder fires phase-locked action potential (meaning, the waveform of the action potential is always the same) at a fixed delayed time after each outside transient is formed. Therefore, time coders neglect information about waveform and amplitude, but focus on frequency of the signal and fire action potential on a 1:1 basis. Amplitude coders, on the contrary, fire according to the EOD amplitude. While both wave-type and pulse-type fishes have amplitude coders, they fire in a different way: receptors of wave-type fishes continuously fire at a rate according to their EOD amplitude; receptors of pulse-type fishes fire burst of spikes to each EOD detected and the number of spikes in each burst is correlated to the amplitude of the EOD. Tuberous electroreceptors show a V-shaped threshold tuning curve (similar to auditory system), which means they are tuned to a particular frequency; the frequency they are tuned to is usually closely matched to their own EOD frequency.

Classification of Tuberous Organs

Type	Fire according to	Method of encoding	Found in
Time Coder	Frequency of received EOD	Fires action potential in a 1:1 ratio to the received EOD	Both types of weakly electric fishes
Amplitude Coder	Amplitude of received EOD	Wave-tyep: continuously fire at a rate according to EOD amplitude/ Pulse-type: number of bursts in each spike is depend on the amplitude of EOD	Both types of weakly electric fishes

Electric Organs

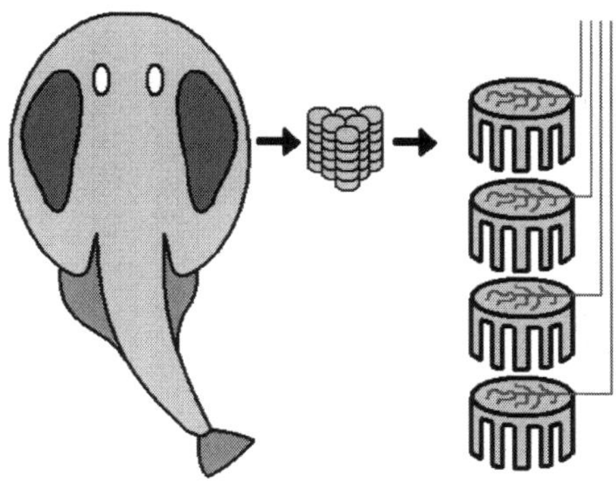

Figure: Electric organ

Weakly electric fish generates Electric Organ Discharge (EOD) with specialised compartments called electric organ. Almost all the weakly electric fishes have electric organs derived from muscle cells (myogenic) ; the only exception is the Apteronotidae, a family under

Gymnotiforms, that has electric organs derived from nerve cells (neurogenic). Myogenic electrocytes are arranged into columns of small, disks-like cells called electroplate. Apteronotidae, being the exception, also carries myogenic electric organs in larval stages; however, as the fish matures, electrogenic organs derived from central spinal cord gradually replace the muscle cell-derived electric cells.

Discharge of an electric organ begins with central command from a medullary pacemaker that determines the *frequency* and *rhythm* of EODs, which is also referred to as SPI- sequence of pulse intervlas; the command is then passed on by spinal electromotor neurons to electrocytes that form the electric organ, which determine the *waveform* of the EODs based on its morphophysiological properties. As commands from pacemaker reaches electric organ, it opens up all the sodium channels, causes net sodium ion flux in one direction (either *towards* or *away from* the head, and brings simultaneous depolarisation of all electroycytes on the same side of cell. The result is a positive polarity at the fish's head relative to the tail, or vice versa: a dipole system. The polarity built up by the electric organ therefore sets up electrostatic fields in the water.

Mormyrids

In mormyrids, the electric organ is fairly small and located only in the caudal peduncle region (the narrow part of a fish's body where the caudal fin is attached to). Electric organ is formed by disk-like electrocytes serially connected together in two columns, and each column resides on one side of the spinal cord.

The myogenic electrocytes are identical to each other and arranged in a homogenous pattern, which results in synchronised discharge-electrical potential recorded from a single electrocyte is equivalent to the miniature version a complete EOD measured outside of the fish. Electrocytes also have an important structure called "stalk," which are tentacle or tube like structures that extend out from each electrocytes. Different stalk-electrocyte systems have been observed, which include stalks that penetrate the elctrocytes, innervate electrocytes from the posterior or anterior, and so on.

Multiple stacks from one electrocyte eventually fuse together to form a large stalk that receives innervations from spinal electromotor neurons. Different morphological structures of the stalk/ electrocyte system result in differences in electric current flow, which further leads to the various waveforms.

Gymnotiforms

In gymnotiforms, electrocytes are quite different between wave-type and pulse-type electric fishes. In wave-type fishes, electrocytes are oriented in longitudinal direction in a tubular form. The electocytes also form columns, but unlike the shorter size of electric organ in mormyrids, gymnotiforms have long electric organs that extend through almost the entire longitudinal body length. Also different from the multiple ways stalk system innervates the electrocytes in mormyrids, stalks in gymnotiforms only make one type of innervations at the posterior side of electrocyte.

Pulse-type gymnotifoms generally show a higher complexity than the wave-type fishes. For instance, their electrocytes can be either cylindrical or drum-shaped with stalks innervating from either posterior or anterior side. Another important difference is that, unlike mormyrids or wave-type gymnotiforms, electrocytes of pulse-type gymnotiforms are not homogenous along the long electric organ that traverses the fish body. The inhomogeneity of electrocytes leads to EODs with highly variable and complex spatial-temporal properties.

Apternotids, being a member of the wave-type gymnotiforms, is different from all other electric fishes by possessing the only neurogenic electrocytes. Electric organ of Apternotids is derived from neurons; more specifically, they are formed from the axons of spinal electromotor neurons. Such structure eliminates one [synapse | synaptic gap] between spinal electromotor neuron and the myogenic electrocytes, which might contribute to Apternotids' highest EOD frequency (>1000Hz) among electric fishes.

Signals

Types of Signals

There are two types of signals generated by electric fishes: pulse-types and wave-types, and each electric fish produces only one of the two distinct types of signals. A pulse-type EOD is characterised by *discrete* EOD fires separated by silent intervals much longer than the discharges; contrarily, a wave-type EOD has its firing period and silence period approximately the same in length, and therefore a *continuous* signal with *quasi-sinusoidal* waveform is formed. Among the two mostly-studied groups of weakly electric fish, the mormyrids and gymnotiforms, both pulse-type and wave-type fishes are found and are grouped by families.

Physical Properties of Signals

Electric Field:

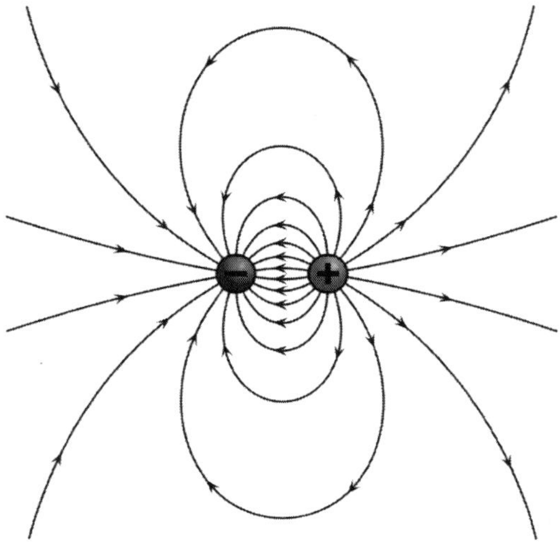

Figure: Electric field built up by a dipole system of magnet. Electric fish generates electric field in the same manner.

Electric fish generates signals by forming an electric field from a dipole-like system made by their electric organ. The electric field from a dipoleforms closed field lines pointing from the positive pole to the negative (can be either head or tail), which makes the electric fish EOD vastly different from other communicating modes that have propagating waves.

While sound waves for acoustic communication, chemical gradients for chemical communications, or light waves (electromagnetic waves) for visual communications all propagate, electric signals do not (it is different from electromagnetic waves). As an electric field, the signal magnitude decreases as the inverse cube of distance (1/(r^3)), which makes the signal sending and formation a *energy-costly* process.

To solve this problem, electric fish matches the impedance of their electric organ to the conductivity of water in order to achieve *minimum energy lost*, and the final result is electric signals travelling for at most a *few metres*. Although electric fish signal has its limit on travelling distance, it does not deteriorate with factors that affect waves, which include reflection, refraction, absorption, interference, and so on. As a result, the temporal features, which are very important for electric fish signals, strictly remain constant in space.

Active Space

When transmitting electric signals in aquatic environment, the physical and chemical nature of the surrounding can make big differences on signal transmission. Environmental factors that might impose influences include solute concentration, temperature, and background electric noise (lightning or artificial facilities), etc. To understand the effectiveness of electric signal transmission, it is necessary to define the term “active space-” the area/ volume within which a signal can elicit responses from other organisms. The active space of an electric fish normally has an ellipsoid shape due to the arrangement of dipoles formed by its electric organs. While both electric communication and electrolocation rely on signals generated by electric organs, electrocommunication has an active space tenfold larger than electrolocation because of the extreme sensitivity of tuberous electrocommunication receptors.

One of the biggest factors that affect active space size will be the *[water conductance]* mediated by solute concentration in the water. It’s been shown that momyrids have adapted its optimal active range in lower-conductivity habitats. One natural phenomenon that supports such theory is that, many species spawn during the time when rivers/ lakes have the lowest conductivity due to heavy rains. Having bigger active space in water with low conductivity will therefore benefit mating and courting. One other explanation tested by Kim and Moller is that, having smaller active space during dry season when mating is not occurring accommodates crowded social spacing without unnecessary signal transmission between individuals.

Frequency and Waveform

Electric fishes communicate with electric signals that possess two main qualities- frequency and waveform. The information in waveform is embedded in the electric organ discharge (EOD) itself, which is determined and fixed by the anatomy and physiology of the electric organ. EOD waveform, in some species, changes with developmental stages. Frequency of EODs and interval duration between them are called sequence of pulse intervals (SPI), which is controlled by the command interneurons in the midbrain and medulla, as stated underelectric organs. Alteration in SPI produces widely varying social signals among electric fish during mating, warning, or identifying. These two properties (waveform/EOD and frequency/SPI) are used by both wave and pulse type fishes for recognition and communication.

EOD Frequency

Frequency is the number of occurrences of a repeating event per unit time. Here, EOD frequency is referred to the firing rate pf an electric fish. Wave-type fishes carry out species recognition by mediating their EOD frequencies, which include their baseline firing frequencies and modulation of frequencies that results in rising, falling, warbling, and cessations of EOD frequencies. For an example, some gymnotiform species use "chirps," a sudden frequency increase, during courtship.

EOD Waveform

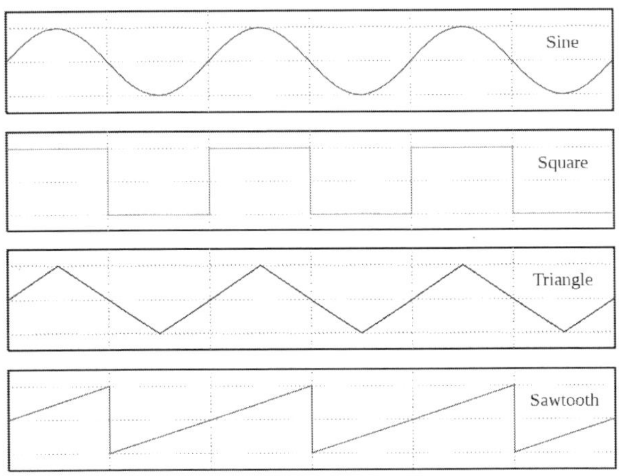

Different waveform representations

Waveform is the shape and form of a wave. Each species of electric fish has their distinct EOD waveform. Pulse-type fishes conduct species recognition by paying attention to the differences of EOD waveform, which include properties such as: EOD duration, number of phases, and form of the phases. Meanwhile, some indirect properties hidden in waveform are also used by pulse-type fishes: amplitude gradient, duration ratios of phases, and the order and signs of phases.

Differences and Changes in Signals

Electric fishes normally possess a baseline frequency and waveform of their signals, alteration in both qualities occur all the time- among different species, sex, development stages, and dominance status. While different alterations occur to signal generations based on the fishes' identities, the level and types of alteration is limited by the fishes' own sensory system, which is biased to sense signals that have similar frequency to its own discharge frequency.

Signals and Sex

As electric fish matures, some taxa develop differences in EOD between male and female (sexual dimorphism). Typically, male electric fish has lower EOD frequency and longer EOD duration than female; among males, the dominated and biggest fish generally possesses the lowest frequency. For an example, as measurements made in Sternopygus marucus (Hagedorn, 1986) showed, males usually generates EOD at about 80 Hz, while females generate EOD at around 150 Hz. Such differences in EOD between the two sexes can be traced back to the changes in the action potential in electrocytes. As electric fish matures from juvenile stage, male fish grows larger with longer and thicker tail, which might result in larger electric organ that generates lower frequency EODs. One of the physiological factors that have been proven to contribute to the sexual dimorphism of EODs is the level of teleost hormone- androgen 11-keto-testosterone (11-KT) and estrogen. Experiments shown that by injecting 11-KT to female electric fishes, not only did their EOD waveforms and frequencies turn closer to that of the males, but their tuberous electroreceptors were also modified to be able to detect signals according to the newly-transformed EOD properties. On the other hand, when estrogen was applied, male electric fishes' EODs gradually turn closer to the female EODs.

Sexual dimorphism in EOD waveforms and frequencies also imposes an influence on active space size. Using species *Sternopygus marucus* as example, male and female emits frequencies almost twice as different from each other (80 Hz vs 150 Hz). However, since most electroreceptors are tuned to signal frequencies that are closer to the receivers' own frequency, the difference in EOD frequency results in different ability of electric fishes to sense signals from either sex, which further leads to different active space sizes. As measured in Sternopygus *marucus* by Hagedorn, male fish can only detect female in a range of *6 cm,* while female fish can detect a male fish in a much bigger range of *39 cm*. This active space size difference is hypothesized to give female fishes a better chance to get closer to potential mates and select the right individual to mate with.

Signals and Development Stages

Studies done on both gymnotiforms and mormyrids have shown that there are species in both groups that have significant EOD changes from larvae to adult. Gymnotiform larvae all have EODs that are simple, monophasicaly similar to a single period cosine function, and formed with a very broad spectrum at lower frequency range. It is observed that, as the larvae mature, the frequency spectrum decreases, the discharge waveform becomes sharper, and more complex waveforms that may consist of multiple phases gradually replace the simple larval EOD.

For the myogoenic fishes, this change in signal waveform occurs with the initial larval electrocytes *fusing* together forming new electrocytes with different shapes, along with redistribution of *ion-gated channels*, formation of new *extracellular structures* on the electrocytes, etc.

Some pulse fishes also develop *accessory electric organs* located on other parts of the body; these extra electric organs further play a role in adding phases to the EODs. For the only neurogenic fish known so far, apteronotids, EOD changes during the developmental process seemed to be more dramatic than that of the myogenic fishes, which might indicate that neurogenic electrocytes are more easily prone to modifications. Similar with the myogenic fishes, apteronotids has its electric organ formed by myocytes. As apteronotids matures, new neurogenic electrocytes derived from spinal motoneurons replace the myogenic electrocytes.

There were two hypotheses proposed for the reason why electric signals were modified during the fishes' development stages. Firstly, as stated above, the fishes' electroreceptors are usually tuned to a specific range of frequencies.

Therefore, to make effective communication, it is necessary for the electric fishes to narrow the broad frequency spectrum of the larval EOD. Secondly, it is known that the electroreceptors of catfish, gymnotiforms, and most pre-teleost fish are tuned to lower frequencies. Therefore, keeping the low frequency of larval EOD will increase the risk of being detected by predators.

Signals and Dominance Status

Measurements have shown that typically, male electric fish that's dominating usually has lower EOD frequency and longer EOD duration. An experiment has shown that, when two males are placed in the same fish tank, both fishes enhance their EOD in the first short period of time. However, after leaving the fishes in a dark period (mimicking night time), the male with higher EOD amplitude, which is also usually the male with bigger body size, will further enhance its EOD;

on the contrary, the male with smaller body size/ smaller EOD does not enhance its EOD.

Special Signals

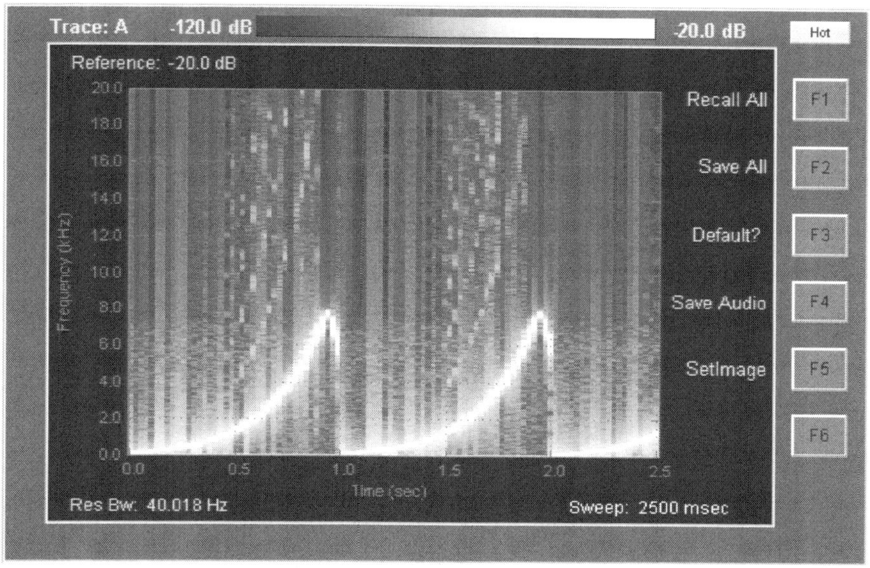

Figure: Spectrogram of a chirp

In electric communication, there are some distinct types of signals that serve special purposes such as courting or aggression. Examples of these special EODs include: "rasps", "chirps" and "smooth acceleration".

Rasp is a burst of pulses at a relatively constant frequency performed by some species during courtship. Chirp is a rapid increase or decrease in frequency.

Smooth acceleration is a period of tens to hundreds of milliseconds that EOD rate increases but in a smooth way. Due to law of conservation of energy, the amplitude of the EOD might lower for a few percent, but the overall changes in waveform and amplitude Is small. Male gymnotiforms emit these accelerated signals during aggression and courtship.

In the fish studied, if courtship goes well and proceeds to spawning, male electric fish starts to use another special type of EOD- the chirp. Chirp also lasts for tens to hundreds milliseconds; however, the increase in frequency was so high that electrocytes could not recover soon enough, and therefore, chirps has a very small amplitude and a waveform deviated from the original waveform.

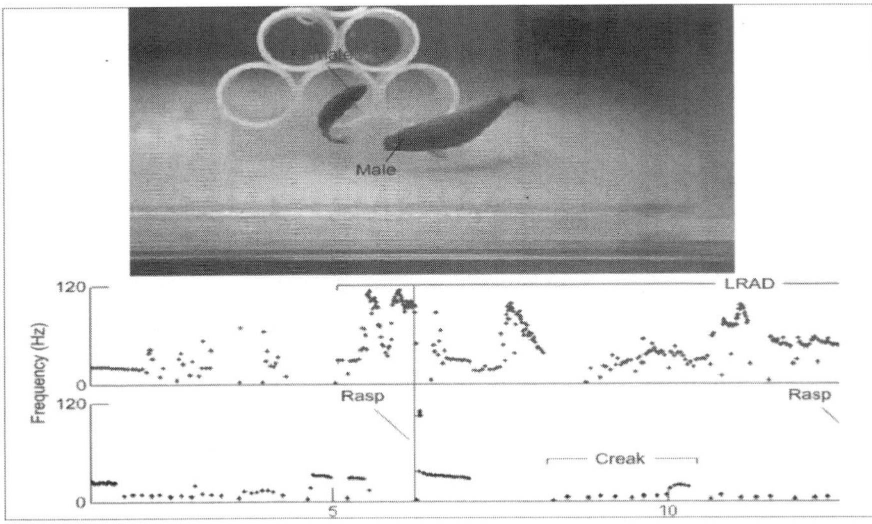

Figure: Courtship and spawning behaviour and EOD signals between a male and female Brienomyrus brachyistius. Shows Rasps and LARD (Long Repeated Acceleration Deceleration) Recorded and Analysed in Carl Hopkins lab.

Measuring Signals

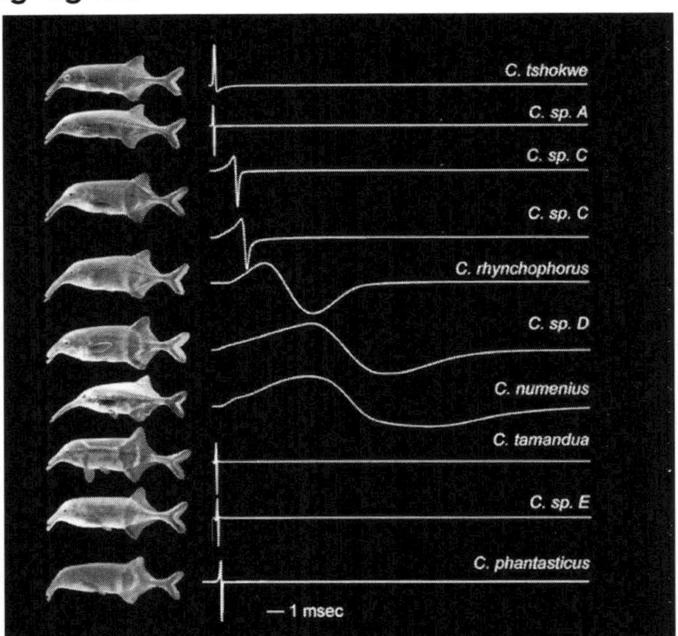

Figure: EODs of different electric fishes. Recorded in South Africa by Carl Hopkins from Cornell University.

Quantitative Measures

Quantitatively measuring and visualising electric signals could mean great challenge and required equipments including electrodes, analogue-to-digital converters, Fourier transforms, Delaunay tessalations, and so on. Measurements and processed-data include electric potential gradients, electric field maps and discharge waveforms/ frequencies. Complex computer simulations, computing, and modelling are also required to further examine the signals.

Coding Electric Signals

In general, neural coding is a process in which information from sensory stimulus, which include light, pressure, and sound, etc., is *coded* into a "language" the brain can understand; these new codes are then represented to the brain by networks of neurons. External stimuli differ on their duration, intensity, form, direction, and more; to represent these differences in sensory stimuli, sensory neurons change their action potential firing in various temporal patterns-sequences, duration, or frequencies.

For electric fishes, EOD coding is the process in which electric fishes code external electric signals, which vary in waveform and frequencies, into patterns of action potential firing and present it to the brain. There are two main EOD coding schemes that have been studied: rate codingand temporal coding. Rate coding focuses simply on the various "firing rate." Temporal code, on the other hand, focuses on the precise timing of single spikes, the timing of rise and falls in a EOD (waveform), etc.

Frequency Coding

Rate coding in the field of neural coding has been the earlier and traditional coding scheme, which assumes that information is only carried in the rate of firings. Due to this reason and our understanding in action potential's all-or-none properties, it was natural that the researches in EOD coding started with frequency coding. Generally, frequency coding is established on the following principles:

- Electric fish has their own firing frequency range
- Frequency normally increases with stronger stimuli
- Combined frequency is used to communication in some wave-type fishes.

Wave-type electric fish uses frequency coding during electro-communication. For one thing, since wave-type fishes *constantly* generates a *continuous* signal, they need a strategy to distinguish *self* and *neighbour* signals. For instance, in order to prevent interference between two signals, when two neighbouring fishes generate signals that have frequency too close to each other, both fish temporarily shifts its frequency away from the original point- the well-studied Jamming Avoidance Respond. Before the actual circuitry and neurophysiological properties were solved, there were several behaviour experiments done on wave-type fishes that proved the usage of frequency coding:

- When playback of artificially generated sine-waves with different frequencies were presented to males, they show greater number of *attacks* and *chirps* when the frequency is within their own range of EOD frequency.
- When playback of male chirps within the conspecific frequency to females, they produce spawning behaviour.

Combined Frequency

Following the behaviour experiments, further researches have shown that wave-type electric fishes utilise characteristic of combined-waves to identify species, sex, and other status of the other fishes. Since wave-type fishes are constantly firing and have no corollary discharge, their tuberous organs receive electric signals from other fishes and their own EOD at the same time. When two signals with very close frequencies (difference < 5Hz) comes in, due to the physical properties of waves, the two waves will "combine" and form two new sinusoidal waves from the modulation of both *frequency* and *amplitude*, which requires the participation of both amplitude-coding and time-coding receptors.

Amplitude modulation results in "beats"- a new sinusoidal wave with its frequency equal to the frequency difference of the original two waves ($fnew = | f1 - f2 |$). Amplitude-coding units read the beats and fire synchronically with the beats. Therefore, amplitude-coding units provide information on the difference in frequency (–f) but cannot determine the *"sign,"* which essentially means "which fish has higher EOD frequency." To determine the sign of –f, the fishes need to compare information from both amplitude and frequency modulation. Frequency modulation of the two signals results in new sinusoidal wave that has firing times alternate between *lagging* and *leading* phases. It is discovered that with different sign of ”f, the relation between phase and amplitude modulation is opposite. (When ”f > 0, phase lags while amplitude increase and leads while amplitude decrease; when ”f < 0, phase lags when amplitude decreases and leads when amplitude increases). Therefore, by integrating information from two sources (self and another sender) and then combining information from two modulations, wave-type fish is able to determine the EOD frequency of another fish and determine the identity of others.

Temporal Coding

Temporal coding was first being neglected by researchers studying electric fish signals because:

1. EOD seemed to be too short to include information within
2. The idea of communicating with signals coded in the same way of action potential was easier to be accepted

Therefore, research on signal coding was first focused on SPI (sequence of pulse intervals) until playback experiments were done in the history to prove that, temporal coding fishes (pulse-type) focus on EOD itself but not SPI. Utilising male fish courting calls of rasps (highly rapid discharge), experiments done by Hopkins that proved the low importance of SPI in pulse-type fish include:

- When playback of sine wave with conspecific female SPI was presented to males, no response was generated from the male (which proved that frequency was not the code)
- Male fishes were presented with playback of 1) original female EOD; 2) phase-shifted female EOD; 3) time reversed female EOD- No response in the phase-shifted group but rasping behaviour was observed in the other two playbacks
- Male fishes were presented with computer generated signals that have the following options mixed up in all possible combinations: male/ female, conspecific/ not-conspecific EOD, and conspecific/ not-conspecific SPI- It was observed that to elicit rasps, the combination must include: female, conspecific EOD, and lastly, either types of SPI

These experiments pointed to a coding scheme that pays attention to EOD itself (waveform, amplitude gradient, duration ratios of phases, etc.) but not the the frequency. Further researches found out the receptor that is responsible for temporal coding: the knollenorgan receptors. When the sender builds up an electric field during communication, electric current flows into the receiver body from one side and exit from the other. Therefore, between electric receptors on each side, the direction of electric current flow is opposite and results in opposite stimulus polarities to the receptor. Electric physiology recording shows that under normal polarity (positive to negative, which occurs on the side where electric current enters the fish body), knollenorgans fire to the onset of a current; on the contrary, under reversed polarity (negative to positive, which occurs on the side where electric current leaves the fish body), knollenorgans fire to the offset of a current. By comparing the inputs from knollenorgans on each side of the fish, a pulse-type electric fish can determine information of the signal that includes waveform, wave duration, and so on.

Colouration in Various Fishes

Visual Appearance

Visual appearance, especially colour, is the most important characteristic of foods in determining their selection prior to actual consumption. "Carotenoids" are of particular importance because they serve as one of nature's most important pigments. In biology, any substance that can impart colour to the tissues or cells of animals or plants can be called a pigment. Carotenoid is the generic name for one of the most common groups of naturally occurring pigments found in the animal and plant kingdoms. The most familiar is beta-carotene, the orange-red pigment in carrots, that was discovered almost 200 years ago. It is from the word carrot, that the group name, "carotenoid" is derived. Carotenoids are synthesized around the globe by plants, fungi and some bacteria at the rate of some 3 tons every second. Carotenoids show great diversity in structure, function and colour. To date, over 600 carotenoids have been identified in nature, varying in colour from yellow to red. Carotenoids are present without exception in photsynthetic tissue and occurring with no definite pattern in nonphotosynthetic tissues such as roots, flower petals, seeds, and fruits. They are also found in fungi, yeast, moulds, mushrooms, and bacteria, and in many cases they are the major pigment in some tissues of aquatic and avian species, that is, lobster, prawn, salmon, and in the feathers of many birds. Most carotenoids are polyunsaturated hydrocarbons, containing 40 carbon atoms, and comprising two terminal ring systems. Carotenoids that are composed entirely of carbon and hydrogen are known as carotenes, while those that contain oxygen are termed xanthophylls.

What Carotenoids do

Carotenoids have many other roles besides colouration. In many cases colouration is not the first role for carotenoids. Research has shown that carotenoids can function as a provitamin A source and as potent antioxidants. Carotenoids have been shown to enhancing the immune response, the may be important in preventing cardiovascular diseases, age-related macular degeneration and cataracts, protection from certain cancers. They have been shown to need for normal reproduction in cattle.

Animals, including humans, cannot make carotenoids from simpler molecules, they are totally dependent on the amounts supplied in their food. In this respect, carotenoids are like vitamins; if intakes are inadequate, health and survival may be compromised.

Colouration in Fish

Colouration plays a role in social structure as well as defence of fish in the wild. Fish depend a great deal on vision as a source of sensory information. All but a few species (mainly cave dwellers) have well-developed eyes. As is the case in many other groups of animals, the body colours of fish are predominantly dependent on the presence of special cells in the skin, called chromatophores. These cells contain pigments or light-scattering or light-reflecting organelles. Carotenoids also have excellent antioxidative characteristics. Cold-water fishes, like salmon, have a high level of polyunsaturated fat in their membranes, and protection of lipid tissue from peroxidation seems to be a metabolic function for astaxanthin. Astaxanthin has been shown to be one hundred times more effective than vitamin E as anantioxidant.

Colouration in Fish

How Fish use Colour

Since Darwin's time we have known that, throughout the animal kingdom, the flashiest males have the best chances of finding a mate. But it was only recently that researchers discovered why this is so. Males that can afford to use carotenoids as ornaments are healthier and stronger. One reason why females are not usually highly coloured is that they need to pass on their carotenoids to their offspring. Carotenoids have been shown to improve young fish survival and growth. Researchers are now pretty sure that carotenoids play a major role in evolution. Carotenoids stimulate the immune system as well as detoxification processes. Animals that are more susceptible to disease and parasites have to use their carotenoids to defend themselves. Genetically resistant males use fewer carotenoids for fighting disease and so can make a better show. Healthier females have more carotenoids for their young. Embryos with adequate stores of carotenoids have a better chance of survival.

It is clear that carotenoid pigmentation of salmon is an important natural indicator of biological superiority. The bright pink colour of salmon flesh and the bright red of salmon eggs are therefore reliable signs of good health in farmed salmon.

Where Fish get their Colour

There are four main groups of pigments that can be used to provide colour in these cells: melanins, carotenoids, pteridines, and purines. Melanins are responsible for the dark colouration seen in fishes. Carotenoids, which are lipid-soluble, dominate in giving the yellow to red colours. Pteridines are water-soluble compounds and result in bright colouration like the carotenoids. Pteridines play a small role in colouration when compared to carotenoids. In the purine compounds, guanine predominates, and large amounts of guanine can be found in the silvery belly skin of most species of fish.

These basic compounds can be combined with other components, like proteins, to produce the blue, violet, and green colour ranges seen in fishes. In the flesh (muscle), the carotenoids are the dominant pigment. In lobsters and shrimps, astaxanthin is attached to a protein to produce the carotenoprotein, crustacyanin. This carotenoprotein imparts a blue colour in living Crustacea; in the presence of heat the carotenoprotein molecule is cleaved, which subsequently results in the characteristic red colour of cooked lobsters and shrimps. Carotenoids are the major pigmenting compounds and cannot typically be synthesised by fish. In contrast, most other pigmenting compounds can be made by the fish.

Which Carotenoids are in Fish

Astaxanthin is the pigment that provides salmon flesh with its characteristic rich pink-red colour and the major carotenoid in the integument and internal organs of red sea bream. Astaxanthin accounts for more than 90% of the total carotenoid content found in the flesh of wild salmonids (salmon and trout). In the wild, fish absorb astaxanthin from the crustaceans they eat. The absorbed carotenoid is then transported in the blood to the muscles and skin where it is deposited. In other fishes and to a limited extent in salmonids, additional carotenoid compounds are the source of other bright colours. Research results indicate that tunaxanthin is a rather common pigment in marine fish. This carotenoid is especially abundant in yellow-coloured fishes, like yellowtail (Seriola quinqueradiata). Astaxanthin, by contrast, seems to be dominant in red marine fish. Lutein is also widely found in many marine species. Carotenoids commonly occurring in freshwater fish include beta-carotene, lutein, taraxanthin, astaxanthin, tunaxanthin, alpha-, beta-doradexanthins, and zeaxanthin.

- Lutein
- Tunaxanthin
- Beta-Carotene
- Doradexanthins
- Zeaxanthin
- Canthaxanthin
- Astaxanthin
- Eichinenone.

Carotenoids in Aquaculture

Carotenoids are used in aquaculture feeds to provide the colour associated with consumer products, such as the bright vibrant colours of ornamental fish. The same carotenoid, astaxanthin, found in wild salmon is used in aquafeeds to impart this natural, pink-red colour to farmed salmon fillets. Colour matters, particularly in regards to consumer preference for aquaculture products.

A Fish of a Different Colour

Figure: *Danio rerio*, the zebrafish

The zebrafish, a small but flashy aquarium pet, may seem like an unlikely informant on questions of human genetics — yet its genome could hold the keys to understanding many diseases and, surprisingly, the genes underlying human skin colour. In December 2005, a cancer research team headed by Keith Cheng at Penn State University announced that their studies of the mutant "golden" zebrafish had taken an unexpected turn: they had discovered a single human gene that accounts for about 30% of the difference in skin colour between African and European descendents. This colour difference can be traced back to a tiny difference in two versions (or alleles) of the gene. The "G" allele is common among African descendents and causes more of the dark pigment melanin to be present in skin cells, while the "A" allele is common among European descendents and causes less melanin in skin cells.

Where's the Evolution?

But how did scientists get from a "golden" fish to human ethnic groups, and what does this all have to do with evolution? The answer relies upon the evolutionary relationship between humans and fish and helps illuminate how and why human skin colour differences evolved in the first place. Though the Human Genome Project has revealed the string of three billion base pairs that represents our genetic sequence, we still know relatively little about the meaning of that sequence. However, because all life on Earth shares common ancestors and because we and other species have inherited many of the same genes from those ancestors, we can learn more about our own genome by studying the genes of other species. For example, researchers can experimentally manipulate a gene in another organism (such as a zebrafish) in order to figure out how the gene works and then look for the homologous gene in humans — the version of the gene that humans also inherited from an ancestor we share with the study organism.

That strategy was the approach that the Penn State researchers took when they began studying the zebrafish. The common ancestor of humans and zebrafish swam in the oceans more than 400 million years ago. Despite the enormous length of time that separates those two lineages, they share many genes, such as this recently-discovered pigmentation gene. The zebrafish and human versions of the gene are about 70% identical. In zebrafish, the gene codes for dark stripes — but by breaking the gene and cutting short the protein that it produces, researchers can engineer a golden-hued zebrafish without dark stripes. The same gene was also discovered in the human genome, where it explains much of the difference in skin colour between African and European descendents. Interestingly, the gene variants that correspond to dark and light skin in humans are remarkably similar to one another — they differ by just one base pair!

But how did different ethnic groups wind up with these different versions of the pigmentation gene? Research suggests that more than

100,000 years ago, the earliest humans lived in Africa and carried the "G" allele, which causes lots of melanin to be present in skin cells and hence, dark skin. Melanin absorbs UV rays from the sun and controls the amount of UV radiation that penetrates our skin. Our bodies need some UV radiation (to build the essential vitamin, vitamin D) but not too much (because UV radiation can damage the skin and destroy another essential vitamin, folate). In the sun-drenched environs of Africa, dark skin was advantageous, preventing UV rays from doing too much damage, while allowing in enough UV to synthesize vitamin D. In that environment, individuals born with a mutant version of the gene associated with less melanin and lighter skin would probably have had poor health and low reproductive success.

Many biologists hypothesize that between 55,000 and 85,000 years ago, humans began to migrate out of Africa. Some of them wound up living in the colder, darker climes of Europe. There, too much UV radiation was not a problem, but too little UV to synthesize vitamin D probably was. At some point either before or after the migration out of Africa, a mutation occurred in one of the ancestors of modern Europeans. This mutation was tiny, changing just a single base, but it caused much less melanin to be present in the skin of those who carried the mutation. This was the "A" allele. Among the new Europeans, this allele likely had an advantage over the "G" allele. Individuals carrying the "A" allele had less melanin, which allowed more UV light to penetrate their skin, which could have allowed them to synthesize vitamin D better than those carrying only the "G" allele. These "A"-carrying individuals had increased reproductive success in their sun-poor environment, and via natural selection, the "A" allele spread throughout the European population. Meanwhile, among Africans, the "G" allele continued to be advantageous and to maintain its majority there.

The discovery of this pigmentation gene has helped us piece together a more complete picture of the evolutionary changes that underlie skin colour differences between human ethnic groups. It suggests that a great deal of the skin colour difference that has delineated the boundaries of social tensions for much of recorded history can be traced back to a tiny genetic change that allowed humans to better survive and reproduce in particular environments.

This discovery also highlights the fundamental similarity of all humans. Skin colour, it turns out, really is skin deep; at a genetic level, the skin colour difference between a European descendent and an African descendent might be largely influenced by a single base pair difference in a genome composed of three billion base pairs. However, neither does this single gene tell the whole story. For example, many East Asians carry the "G" allele and yet have light-coloured skin. Why? We don't know, but it is probably due to many other undiscovered genes. And though, on average, people who carry only the "A" allele have lighter skin than those who carry only the "G" allele, there is a huge overlap in skin colour between the groups — so knowing which alleles an individual carries will not let you predict with much certainty the shade of his or her skin. It is clear that the concept of "race" is more a sociological construct than a biological one, and that many genes and environmental factors influence skin colour. As we learn more about these genes, we will learn more about our own evolutionary history as a species.

Finally, methods used to discover this pigmentation gene reveal the deep connections between our own species and others that have lived over the approximately 3.5 billion years of life's history. Studying the zebrafish, for example, led to discoveries about our own evolutionary history and genetic variation. And because of the common ancestry of all life on Earth, other species (as different from us as they may seem) represent our evolutionary cousins and can teach us about ourselves and our history.

Chapter 7

The Physiology and Behaviour of Colour in Fishes

Long before aquariums we've known that fishes change colour in response to their background, and that they change colour during exercise and courtship. These changes in appearance are under the control of pigment containing cells called "chromatophores". These are the same types of cells present in some invertebrates: crustaceans, insects, cephalopods, and are found in other cold-blooded vertebrates (reptiles, amphibians). The first to review the literature on the physiology and behaviour of animal colour (Animal Colour Changes) was Parker (1948). This work was further compiled and added to by Bagnara and Hadley (1973) in the third volume of Fish Physiology: Chromatophores and Pigments. Importantly this last work covers how fishes use their chromatophores in a variety of responses to internal and external stimuli.

Physiology of Colour, Change

Delving through even recent literature one will find some disagreement re defined terms, but the following terms are generally agreed upon. A single colour cell is a chromatophores, a group of chromatophores a chromatosome. Chromatophores are defined on the basis of the pigments they contain:

Principal Types of Chromatophores and Properties

Chromatophore	Pigment	Type	Colors
Melanophore	Melanins	Color	Black, Brown
Xanthophore	Pteridines and Carotenoids	Color	Red, Orange
Erythrophore	Pteridines and Carotenoids	Color	Orange, Yellow
Iridophore	Guanine and other Purines	Reflective	White, Silver, Blue, Others

Types of Chromatophores

Colour and its change in fishes is a matter of colours already present being manifest, clumped or masked. Of the most common types of chromatophores are melanophores. These contain black or brown coloured melanin crystals. The degree or intensity of darkness of these cells and hence the fish they're found on depends on the amount of dispersion of the melanin pigment within the cell. Intracellular migration and aggregation is the fundamental chromatophores activity. When the pigment is dispersed widely throughout the cytoplasm, the skin macroscopically appears darker (black or brown). When the pigment granules aggregate or contract the cell loses its darkness.

Some examples of this loss of dark colour are the dark bands on freshwater angels (*Pterophyllum*) and sunfishes (*Lepomis*) that can quickly blanch from dark to light or come back again given fright or excited states. Other coloured cell types include xanthophores and erythrophores. These chromatophores are generally slower to change. An example of these cells can be understood readily by looking at a swordtail (*Xiphophorus*).

Studies of skin pigment cells in recent years have helped our understanding of pigment biology. Two important concepts have become established re the function of chromatophores re the dermal unit and the epidermal melanin unit. The epidermal unit is only concerned with morphological colour change (in warm and cold-blooded animals). The dermal chromatophores unit contains three layers:

- The xanthophore or filtering layer
- The iridophore or reflecting layer
- The melanophore or absorbing layer.

Note that there are two types of pigments true or based on colour and reflective. Also note that not all chromatophores contain the actual pigment colour that they appear. That is, that some work on different principles other than selective absorption/reflection. The iridophores contain quanine crystals that reflect different wavelengths of light, which give them an apparent colour though no true pigment is present. Further, there are two types of iridophores or reflecting pigment cells, ones with decidedly larger and smaller quanine crystals. The larger crystals can change their orientation to reflect different colours of light. Cells with the smaller crystals can aggregate or disperse their pigments thereby controlling the intensity of colour.

Because iridophores are typically light in colour, the effect of dispersion and contraction is opposite that of melanophores. When quanine crystals are aggregated, the cell appears darker. The plate-like crystals give off iridescence as seen on the top and flanks of many fishes (e.g. Silver Dollars, *Metynnis, Mylossoma*, Anchovies, *Engraulis*). Green, gold, red, blue and many other colours can be reflected selectively by iridophores. The iridescent blues of Neon Tetras (*Paracheirodon innesi*) are a result of quanine crystals for instance, and not blue pigmentation.

Fishes also utilise combinations of pigment cell types, with iridophores and melanophores mixed.

Chromatophore Responses: Physiological (Nervous, Hormonal) and Morphological

Chromatophore changes can be divided into two categories, morphological and physiological. Morphological changes are usually evoked by maintaining an organism in a given setting, on a specific background for a number of days. We'll discuss these states a bit later.

Physiological colour changes involve alteration of pigment granules causing dispersion or aggregation consequent to various stimuli, e.g. light, temperature, chasing.

The control of aggregating and dispersing of pigment granules is caused by changes in the chromatophores ionic charge. A change of charge within the cell causes a change in colour. There are two ways to change the ionic equilibrium within chromatophores, hormonal and neural. Both "paths" are often employed, one working more gradually, the other more immediately. For example, the time required to change from light to dark varies immensely. This change can take from scant seconds in the Killifish Fundulus, to about twenty days in the European Eel (Anguilla). The significance of the different time intervals in colour change is that it is related to the mechanism of change. When the time for the change exceeds two hours, hormonal coordination is indicated. Alternatively, a change in the span of ten minutes of less is likely largely due to neuron-control.

Physiological responses are defined as either primary or secondary and as stated are moderated by nervous and endocrine system control. Primary colour changes generally occur through routes other than the eyes, i.e. by direct action of light on the chromatophores themselves (known as extraocular reflex). Primary responses are functional before the eyes are operative in embryonic and larval stages. Secondary responses depend on the nature of the environmental background and not specifically on the intensity of ambient light. Here the degree of pigment dispersion is determined by the ratio of the amount of light directly incidental to the eye to the quantity and quality reflecting on the eye from the background.

Though colour and its change are mediated physiologically by the nervous and endocrine systems, they are determined by direct environmental influences like pH, salinity, and temperature dispersing pigments without the fish's direct control. Hence a third type (morphological) of colour and pattern control exists in addition to hormonal and nervous.

Morphological Colour Changes

Morphological colour changes are due to amounts of pigment present in the chromatophores of an organism. Morphological changes occur very slowly, generally over the course of a month or more, and are usually permanent. These changes involve the synthesis and destruction of large amounts of pigment through a few methods:

- More or less pigment within the cells
- More or less chromatophores themselves
- The movement of pigment to different layers in the epidermis.

More often than not, morphological colour changes are actionable with the other types of colour change. For example, a fish placed on a white background will contract the pigments in its cells first physiologically (under nervous impetus), and then slowly lose its morphological colour.

Research has shown that the pituitary gland, through its production of melanophore stimulating hormone (MSH) participates in morphological colour changes. Further, that if an animal is sensitive to both physiological and morphological changes by the injection of MSH then the morphological process will be slower and more permanent, perhaps irreversible.

Nervous/Neural Mechanisms of Chromatophore Control

There is good evidence that melanophore control by advanced bony fishes is principally actuated by the autonomic nervous system. Sharks and other cartilaginous fishes colours are apparently not controlled by nervous mechanisms, and consequently do not change rapidly. There are two principal chemicals that are produced and release by neurons (neurohormones) that affect colour.

Epinephrine (Adrenalin): A nerve-activated hormone that's released by an organism when it is excited or scared, causing pigments to contract and the animal to blanch, lose colour.

Acetylcholine: A chemical that is active in muscle tissue, movement, almost always causing melanin to disperse, darkening the organism.

Hormonal Mechanisms of Chromatophore Control

Hormonal colour changes are slower, harder to observe and more subtle typically. Several "chemical messengers" (hormones) affect colour. In fishes, all these are carried by the blood system (endocrine means "to cry within"). Most arise fro the pituitary, although they can be produced in the hypothalamus, pineal, thyroid and other endocrine glands. A comment for those unfamiliar with the structure and physiology of the "lower vertebrates" called fishes: they share about the same organs, chemicals and pathways as "higher ones" like ourselves.

For sure the "less advanced fishes", the Lampreys and Hagfishes and Cartilaginous fishes (sharks, rays, skates, chimaeras) do not have neurological colour control. Instead MSH produced by their pituitaries is responsible of pigments. "Higher fishes" (teleosts) have a slightly different structure to their MSH, resembling that of mammals. Melanophore Stimulating Hormone: The most abundant and studied, though not the only hormone affecting colour change in fishes is MSH (or Intermedin in older literature). A similar hormone MCH, Melanophore Concentrating Hormone has the opposing effect of lightening individuals.

Estrogen: We'll mention as many fishes exhibit "sexual dichromatism" or difference in colour associated with sex due to this hormone and other sex hormones. Wrasses (family Labridae) are good examples here. Other sexual hormones of fishes include Methyl Testosterone, Follicle Stimulating Hormone (FSH) and Luteinizing Hormone (LH). These all influence colour in fishes though not as pronounced or widespread as estrogen.

Functional and Behavioural Significance of Colour, Changes

Behaviour of Colour

Apart from its physiology, the study of the function of colour in fishes is a rapidly expanding scientific field of study. As presently understood and classified, colour falls into three general categories of function:

- Camouflage (concealment)
- Advertising
- Disguise.

Of course, there are other ways to classify colour. For instance a classification scheme might measure the amounts and variability of changes therein. One group here would be those fishes with a fairly stable colour pattern, like the jawless fishes and sharks. Though there are sharks (e.g. the genus Carcharhinus) that show colour change diurnally. Fish with numerous iridophores often do not exhibit elaborate colour changes (they're "shiny" all the time). The Tinfoil Barb (*Barbonymus schwanenfeldii*) is a good example here.

Another scheme of arranging fishes by colour would be by how they do so with their behaviour. Most fishes do change colour by one or more of the methods discussed earlier.

Camouflage

This is the most widespread and important type of colouration. This really makes sense when you consider the importance of not being seen to avoid being eaten. Camouflage colouration is accomplished in a few ways.

a) The most common type of colouration effect occurs in most, but particularly pelagic fishes, counter-shading. This is the matching of reflectivity from below and above with ambient reflection. Melanophores cover the topside of the fish, making it darker, and iridophores start about halfway down their sides of the body, increasing in density until almost solid underneath the fish. Next time you see a show with tunas, jacks swimming in the open sea, or visit a public aquarium displaying herrings or anchovies, look for this.

b) Deepwater fishes have bizarre and unusual colouration. Most are black or red, with the red usually being very bright. As the red-end spectra of sunlight are the first to be selectively absorbed in the upper water of the sea this red colouration is a mystery. It could be that the colour is in some way related to storage of oxygen.

c) Cavefish of a few species live in worlds w/o light and have lost their pigment through evolution. Most have lost their eyes as well and use other senses to navigate. The blind Cave Tetra (*Astyanax jordani*) is a fish sold in the aquarium trade that is almost devoid of pigment and totally blind.

d) Matching camouflage is extremely widespread, especially in the tropics. There are several excellent examples of fishes that have active colour patterns that match their immediate environment. Some fishes even have a repertoire of several adaptive colour patterns. George Barlow (1981) recognised up to eleven such colour patterns in *Badis badis*. Most of these patterns were found to match different surroundings.

Another classic example for pet-fish keepers is the pencilfish *Nannostomus beckfordi* that has a twenty-four hour rhythm of day and night colouration. A horizontal band runs the length of the body during the day while at night three vertical bands are visible to match the water plants it hides in during these hours. It was found (Abbott 1973) that this fish did not depend on light intensity or vision at all and that blind fish still went through the diurnal matching periods.

The vertical bands of freshwater angelfishes (*Pterophyllum*) also match vertical bands of plants, and lighten and darken with the presence of such physical objects.

e) Transparency. One of the more clever forms of concealment is to be transparent. Indian Glassfish (*Parambassis ranga*) and Glass Catfish (*Kryptopterus bicirrhis*) are two good examples of fishes that use this scheme that are common in the ornamental fish trade. This is obviously a good strategy for avoiding predators that rely on sight.

f) Another widespread colour scheme is termed disruptive colouration. These markings serve to distract a predator, the fish thus creating a diversion. False eye spots as in many of the marine butterflyfishes (Chaetodontidae) or the Oscar (Astronotus ocellatus) are examples of disruptive colouration, potential predators either scared away or attacking this area allowing the would-be prey to escape. There are many tropical marine fishes with types of disruptive colouration.

Advertising

Advertising is another function of colour and colour changes in fishes. In such large and sometimes dark, always relatively dense environments as the waters of the world advertising ones presence, species identity and possibly sex are critical.

a) An excellent example of bold advertising off the U.S. California coast is the Garibaldi (*Hypsypops rubicunda*). The Garibaldi uses its bright orange colouring as an advertisement to others of its own kind to defend its territory. These territorial boundaries are set by sight... becoming smaller in more turbid water, and larger in more clear water.

b) Recognition is a very important function of colouration. Both pattern and colour are often used in different types of recognition.

Species recognition is important to fish as it is to us. People often use colour and patterns to recognise different species. Though not touted by systematists for discernment, humans being visually oriented often rely on colour and markings for species identifications.

Sex recognition. Though not obvious to most of us, there are often morphological and/or colouration differences between the sexes of fishes. Often this brightened condition is disadvantageous during non-breeding seasons; hencefishes show their most heightened colouration only while attracting mates and breeding.

A third type of recognition is warning colouration, within or between species. Changing colour to warn of impending trouble.

c) Another example of advertising involves colour and cleaning symbiosis. Cleaner fishes (e.g. *Labroides* wrasses) are brightly coloured so they will attract largerfishes that want to be cleaned. This "signal releaser" colouration also advertises that the cleaner is not to be ingested as prey. Some other species of fishes utilise this protection (as mimics) like *Aspidontus*, the saber-tooth blenny.

Disguise: Mimicry

This is the third major type of colouration and change used by fishes as means for survival. Some species of fishes utilise the looks (and behaviour) of others as a means to "get by" or get a meal. An example is a mimic of the common Labroides wrasse *Labroides dimidiatus*, *Aspidontus*, the Sabertooth blenny, that uses matching colouration for protection from predators and to gain proximity to hosts as food.

Other Uses and Responses of Colouration:

There are several "secondary" apparent uses, results of colouration and its change:

1) Thermoregulation. Some fishes are notably darker in the early morning hours, lightening up with increased temperature, even with loss or constant illumination, and darkening on one side if the water is warmed there.

2) Environmental Responses. Most fishes have the capacity to adapt to some extent to changes in the shade, colour or pattern of their environment.

a) Response to background has been widely observed. The best example here are some of the flatfishes (Pleuronectiforms) that can adjust their body markings, even to checkerboard, to match the substrate. A curious observation is that "practice" shortens the span of time for these matching.

b) Response to darkness. This is a varying of colouration based on direct illumination on the animal.

3) "Psychic Response". Reactions by fishes from being handled or darkening when angry (think of a red devil cichlid (Amphilophus) or triggerfish (Balistidae) by what it observes outside.)

Cloze

Control of colouration has been found to be a matter of the presence and concentration of pigment and reflective elements in types of cells called chromatophores in fishes. These colours and their change are under environmental, hormonal and neural control. There are differing classification schemes for lumping/splitting colour functions and means, but there is no doubt that such markings and abilities to change are of great survival value and multiple functions.

Migration in Fishes

Many types of fish migrate on a regular basis, on time scales ranging from daily to annually or longer, and over distances ranging from a few metres to thousands of kilometres. Fish usually migrate because of diet or reproductive needs, although in some cases the reason for migration remains unknown.

Classification

Classifications can be either fundamental (like biological classification that rests on a phylogenetic basis), or are merely heuristic typologies (as here, about migrations) to assist communication about complex issues. Classifications are judged according to their fundamental accuracy, whether they are convenient or not. Typologies in contrast are essentially arbitrary and their effectiveness is to be judged solely by the problems they solve or create. Secor and Kerr (2009) for example show several typologies that encapsulate various aspects of fish life history.

Migration is a word used in multiple senses. It is important to distinguish "true" migration, i.e. a life-history-structured or at least patterned activity such as seen in *anadromous* species like salmons, from mere movement or wandering as may happen, say, with *euryhaline* species that easily move between fresh and salt water but not necessarily or with regularity. "Vertical migration", for example, the phenomenon of plankton and fishes regularly changing their depth throughout the 24h day, is a special usage unlike migrations of (e.g.) salmons that range over distances in migrations that may cover river, lake, and sea, or the great migrations of game through Africa's Serengeti.

Anadromous and *catadromous* are words that have been commonly used for centuries. They are slightly more narrowly used in the following classification of truly migrating fish by Myers 1949:

"*Diadromous.* Truly migratory fishes which migrate between the sea and fresh water. There has been no English term by which one can refer collectively and briefly to anadromous, catadromous and other fishes which truly migrate between fresh and salt water, and this new term is now proposed. Like the two well known ones, this adjective is formed from classical Greek ([dia], through; and [dromous], running).

Anadromous. Diadromous fishes which spend most of their lives in the sea and migrate to fresh water to breed (From [ana], up). [This narrowed the previous usage which could apply to fish never crossing the freswater/sea boundary but simply moving upstream to spawn, for example in some fishes of the Rift Lakes]

Catadromous. Diadromous fishes which spend most of their lives in fresh water and migrate to the sea to breed. (From cata, down) [This narrowed the previous usage which could apply to fish never reaching the sea but moving downstream to spawn]

Amphidromous. Diadromous fishes whose migration from fresh water to the seas, or vice versa, is not for the purpose of breeding, but occurs regularly at some other stage of the life cycle. (From amphi, around, on both sides).

Potamodromous. Truly migratory fishes whose migrations occur wholly within freshwater. (From [potamos], river). [Rarely used. This term and *oceanodromous* received the fishes excluded by the narrowing of *anadromous* and *catadromous*]

Oceanodromous. Truly migratory fishes which live and migrate wholly in the sea. (From [oceanos], the ocean) [Rarely used. This term and *potamodromous* received the fishes excluded by the narrowing of *anadromous* and *catadromous*] "

Theous endings are for the adjectival form of the terms; nouns are obtained by replacing that ending with "-y", e.g. *anadromy.*

The terms *anadromous* and *catadromous* were of long standing (and similar but not identical usage); the other terms were coined by Myers. Myers was hesitant about introducing new terms however, saying:

> *"The writer's strong aversion to the infliction of new scientific terms upon the scientific public has, however, caused him to proceed with great circumspection. Finally, the writer has ventured to propose new terms only because he feels that certain new terms will be of distinct advantage in some specialised types of ichthyological work, and that the general biologist and fishery worker will seldom or never have to bother his head about them. Catadromous and anadromous will almost certainly always remain the most important and probably* the only widely used *terms of their class."*

Myers' term *diadromous* has proved useful as an inclusive term. But his prediction was otherwise accurate: *anadromous* and *catadromous*, are indeed the main widely understood terms. Secor and Kerr (2009) found only one paper that used *oceanodromous*, 18 using *potamodromous*, and 122 using *amphidromous* compared to 985 times for *anadromous*, 143 times for*diadromous*, 71 times for *catadromous*. Obviously the amount of work done on a group is one source of this variation, but another source is the perceived utility and standing of the term itself.

Amphidromy is rarely understood by the wider fish biology audience, or alternatively "There seems to be some reluctance to use the term amphidromy". Myers' definition's ambiguity ("or vice versa") and teleology ("for the purpose of") is one part of that problem. Its meaning/usage has been adjusted (but two such adjustments were found zero times by Secor and Kerr (2009)), and recently it has been stated as differing from *anadromy* simply because return to fresh water occurs at a juvenile or immature stage; other differences have been claimed (such as where most growth occurs) but they are necessary consequences of the stage at return and therefore are not independent. Some authors have therefore eschewed amphidromy in favour of the more widely understood terms: either *anadromy*, with or without a remark that return to rivers occurs at an earlier stage (e.g. "juvenile-return anadromy"), or *diadromy* which discards information that would be conveyed by *anadromy* with or without remark.

It has to be borne in mind that any typological system has heuristic value as a convenience for description, and need not reflect phylogeny:

each category (term) may include representatives of many distantly related taxa, each of which may well have close relatives that are in another category (term) or that are not migratory and thus fall completely outside the typology. These terms are therefore of limited safety when used alone to screen data to be considered for an analysis, or to decide which species should be read about to explore a phenomenon or possible comparison.

The limitations of typologies were clearly stated by Myers himself. Each may be useful in one context, for one purpose, but not another. I.e. they may categorise along different axes. For example, Myers in the same year devised another interesting set of terms, also directed at fishes, but on the basis of their salt-tolerance.

As a footnote it is interesting to note that George S. Myers had previously been one of the degree supervisors for Porfirio Manacop, whose ground-breaking Master's work (in 1941, later published 1953) on a *Sicyopterus* species overturned the prevailing notion that it was catadromous. It is likely that Myers was impressed by this, and used *Sicydium* as the "type" genus for *amphidromous*.

And, as a contrary footnote showing how science fails: regrettably, Manacop's work seems to have been as locally unpopular as it was innovative, because over 20 years later the group he worked on was still being referred to by a colleague in the Philippines who would certainly have known of his work, incorrectly and without evidence, as *catadromous*.

And although these systems were originated for fishes, they are in principle applicable to any organism.

Forage Fish

Forage fish often make great migrations between their spawning, feeding and nursery grounds. Schools of a particular stock usually travel in a triangle between these grounds. For example, one stock of herrings have their spawning ground in southern Norway, their feeding ground in Iceland, and their nursery ground in northern Norway. Wide triangular journeys such as these may be important because forage fish, when feeding, cannot distinguish their own offspring.

Capelin are a forage fish of the smelt family found in the Atlantic and Arctic oceans. In summer, they graze on dense swarms of plankton at the edge of the ice shelf. Larger capelin also eat krill and other crustaceans. The capelin move inshore in large schools to spawn and migrate in spring and summer to feed in plankton rich areas between Iceland, Greenland, and Jan Mayen. The migration is affected by ocean currents. Around Iceland maturing capelin make large northward feeding migrations in spring and summer. The return migration takes place in September to November. The spawning migration starts north of Iceland in December or January. The diagram on the right shows the main spawning grounds and larval drift routes. Capelin on the way to feeding grounds is coloured green, capelin on the way back is blue, and the breeding grounds are red.

Highly Migratory Species

The term highly migratory species (HMS) has its origins in Article 64 of the United Nations Convention on the Law of the Sea (UNCLOS). The Convention does not provide an operational definition of the term, but in an annex (UNCLOS Annex 1) lists the species considered highly migratory by parties to the Convention. The list includes: tuna and tuna-like species (albacore, bluefin, bigeye tuna, skip jack, yellowfin, blackfin, little tunny, southern bluefin and bullet), pomfret, marlin, sailfish, swordfish, saury and ocean going sharks, dolphins and other cetaceans.

These high trophic oceanodromous species undertake migrations of significant but variable distances across oceans for feeding, often on forage fish, or reproduction, and also have wide geographic distributions. Thus, these species are found both inside the 200 mile exclusive economic zones and in the high seas outside these zones. They are pelagic species, which means they mostly live in the open ocean and do not live near the sea floor, although they may spend part of their life cycle in nearshore waters.

Highly migratory species can be compared with straddling stock and transboundary stock. Straddling stock range both within an EEZ as well as in the high seas. Transboundary stock range in the EEZs of at least two countries. A stock can be both transboundary and straddling.

Other Examples

Some of the best-known anadromous fish are the six species of Pacific salmon, which are Chinook (King), Coho (Silver), Sockeye (Red), Chum (Dog), Pink (Humpback), and Cherry. The salmon hatch in small freshwater streams. From there they migrate to the sea to mature, living there for two to six years. When mature, the salmon return to the same streams where they were hatched to spawn. Salmon are capable of going hundreds of kilometres upriver, and humans must install fish ladders in dams to enable the salmon to get past. Other examples of anadromous fishes are sea trout, three-spined stickleback, and shad.

The most remarkable catadromous fishes are freshwater eels of genus *Anguilla*, whose larvae drift from swanning grounds in the Sargasso sea, sometimes for months or years, before entering freshwater river and streams as glass eels or elvers.

An example of a euryhaline species is the Bull shark, which lives in Lake Nicaragua of Central America and the Zambezi River of Africa. Both these habitats are fresh water, yet Bull sharks will also migrate to and from the ocean. Specifically, Lake Nicaragua Bull sharks migrate to the Atlantic Ocean and Zambezi Bull sharks migrate to the Indian Ocean. Diel vertical migration is a common behaviour; many marine species move to the surface at night to feed, then return to the depths during daytime.

A number of large marine fishes, such as the tuna, migrate north and south annually, following temperature variations in the ocean. These are of great importance to fisheries.

Freshwater fish migrations are usually shorter, typically from lake to stream or vice versa, for spawning purposes. However, potamodromous migrations of Colorado pikemin now of the Colorado River system can be extensive. Migrations to natal spawning grounds easily be 100 km, with maximum distances of 300 km reported from radio tagging studies.

Historic Exploitation

Since prehistoric times humans have exploited certain anadromous fishes during their migrations into freshwater streams, when they are more vulnerable to capture. Societies dating to the Millingstone Horizon are known which exploited the anadromous fishery of Morro Creek and other Pacific coast estuaries. In Nevada the Paiute tribe has harvested migrating Lahontan cutthroat trout along the Truckee River since prehistoric times. This fishing practice continues to current times, and the U.S. Environmental Protection Agency has supported research to assure the water quality in the Truckee can support suitable populations of the Lahontan cutthroat trout.

Modelling Fish Migration

In a paper published in 2009, researchers from Iceland recount their application of an interacting particle model to the capelin stock around Iceland, successfully predicting the spawning migration route for 2008.

Migration in Fishes

Migration is the movement of large number of animals from one place to another for feeding, reproduction or to escape weather extremes. When large numbers of fishes come together and move socially it is called shoaling. But sometimes migrating fishes exhibit high degree of coordination in their movements and carry out synchronised manoeuvres to produce different types of shapes. This is called schooling, as seen in tunas and sardines. Feeding or alimental migration takes place in fishes for feeding. In high populations fishes exhaust food resources in an area quickly and therefore must migrate constantly in search of new feeding resources. Salmons, cods and sword fish constantly migrate for food from one place to another in the sea.

Spawning migration takes place in breeding season in those fishes which have spawning grounds far away from feeding places. Migratory fishes such as eels and salmons and a large number of riverine fishes spawn in tributaries of river in hills and migrate in large number for laying eggs in these oxygen rich waters. Juvenile migration involves larval stages of fishes which hatch in spawning grounds and must migrate long distances in order to reach the feeding habitats of their parents. Recruitment migration takes place when large number of larvae moves from nursery habitat to the habitat of adults which may sometimes be distinctly different. Adults of eels live in rivers in Europe and America but their larval stages live and grown in sea and migrate to reach rivers which may take one to two years.

Seasonal migration takes place in fishes that inhabit arctic areas where in summer climate is conducive and food abundant but as winter approaches temperatures fall below zero and food becomes scarce. Hence fishes must migrate towards subtropical and tropical areas to escape extremes of weather conditions.

Types of Migration in Fishes

Fishes live in two different types of aquatic habitats, namely, freshwater and marine habitats, which pose different osmotic problems because of which it is difficult to migrate from one type of habitat to another. Nevertheless, some fishes do migrate.

Potamodromous Migration

When fishes migrate from one freshwater habitat to another in search of food or for spawning, it is called potamodromous migration. There are about 8,000 known species that migrate within lakes and rivers, generally for food on daily basis as the availability of food differs from place to place and from season to season. Fishes also must migrate to lay their eggs in places where oxygen concentration in water is more and where there is abundance of food for juveniles when they hatch from eggs.

Oceanodromous Migration

This migration is from sea water to sea water. There are no barriers within the sea and fishes have learned to migrate in order to take advantage of favourable conditions wherever they occur. Thus there are about 12,000 marine species that regularly migrate within sea water. Herrings, sardines, mackerels, cods, roaches and tunas migrate in large numbers in search of food by way of shoaling (migrating together socially but without much coordination) or schooling (swimming with high degree of coordination and synchronised manoeuvres).

Diadromous Migration

When fishes can migrate from fresh water to sea or from sea to fresh water, it is called diadromous migration. There are about 120 species of fishes that are capable of overcoming osmotic barriers and migrate in these two different types of habitats. This migration is of three types.

Catadromous Migration

This type of migration involves movement of large number of individuals from fresh water to sea water, generally for spawning as happens in the case of eels *(Anguilla)* inhabiting European and North American rivers.

Both European eel *(Anguilla anguilla* or *Anguilla vulgaris)* and the American eel *(Anguilla rostrata)* migrate from the continental rivers to Sargasso Sea off Bermuda in south Atlantic for spawning, crossing Atlantic Ocean during the journey and covering a distance of about 5,600 km. The adult eels that inhabit rivers are about a metre long, yellow in colour and spend 8-15 years feeding and growing. Before migration the following changes take place in their bodies:

- They deposit large amount of fat in their bodies which serves as reserve food during the long journey to Sargasso Sea.
- Colour changes from yellow to metallic silvery grey.
- Digestive tract shrinks and feeding stops.
- Eyes are enlarged and vision sharpens. Other sensory organs also become sensitive.

- Skin becomes respiratory.
- Gonads get matured and enlarged.
- They become restless and develop strong urge to migrate in groups.

They migrate through the rivers and reach coastal areas of the sea where they are joined by the males and then together they swim in large numbers, reaching Sargasso Sea in about two months. They spawn and die. Each female lays about 20 million eggs which are soon fertilized by males.

First clue about life cycle of eels was given by two Italian scientists Grassi & Calandruccio in 1896. Details of migration and life cycle were later studied by Johann Schmidt (1905). Eggs hatch into leaf-like, semitransparent, larvae having small head called Leptocephalus. Leptocephali of American eels take about 10 months to fully grow while those of European eels take about 18 months. Upon reaching coastal waters leptocephali metamorphose into another larval stage called Elver or Glass eel. Female elvers ascend to the rivers and metamorphose into yellow-coloured adults, while males stay back in the river mouth and wait for the females to return for spawning journey.

Anadromous Migration

Adults of anadromous fishes live and feed in ocean waters but their spawning grounds lie in the tributaries of rivers. Salmons, sturgeons, Hilsa and lampreys are some of the marine fishes that undertake anadromous migration to spawn in rivers.

Atlantic salmon *(Salmo salar)* migrates to the North American rivers for spawning while six species of Pacific salmon (Onchorhynchus) migrate to various rivers of Asian countries.

Salmons living in sea are metallic silvery grey in colour but before migration they turn reddish-brown in colour. During fall, they enter rivers and swim energetically against water currents (contranatent), clearing all obstacles, including waterfalls and reach tributaries in hilly areas where they make a saucer-like pit in which female lays eggs and male releases smelt over them. Eggs take 2-3 months to hatch in the following spring, when the juvenile stage called Alvin emerges out but remains within the nest, obtaining its nourishment from the yolk sac attached to its belly. Alvin then transforms into Fry which feed on planktons. Fries are denatant (they swim along with water current) and feed and grow into fingerlings which take the shape of adult fish. They change into Smolt which congregate at the river mouth in large numbers and then enter sea water in to metamorphose into adult salmons.

Problem of Navigation

How fishes find their way in huge expanses of sea and reach their destinations which lie thousands of kilometres away has been a mystery. It is believed that they orient by the positions of stars and moon in the night sky and sun in daytime to find the direction of swimming. However, it has been experimentally proven by A.S. Hasler that salmons are guided by the odour of their parent stream during return journey. Eels can also migrate to Sargasso Sea using similar odour maps but how leptocephali find their way back to the river mouths, crossing vast stretches of Atlantic Sea is a mystery.

Anadromous Migration in Lampreys

Adult lampreys are parasitic on other fishes and live in sea for 3-4 years and grow to become 30 cm long. For breeding, they stop feeding and migrate in rivers hundreds of miles upstream. Males make nest in sand and gravel in which female lays eggs and male fertilizes them. In about 3 weeks time, eggs hatch into 7 cm long, yellowish-brown ammocoete larva that lies buried in sand and feeds on detritus by filter feeding method. Larva lives in river for 3-7 years and grows from half centimetre to 17 centimetres long. Then these fully grown larvae start their downstream journey and enter the sea to metamorphose into adults.

Chapter 8

Immunology of Fishes

Ecology of Fishes

Feeding

Figure: The bumblebee cichlid, Pseudotropheus *crabro*, is specialised in feeding on parasites from the catfish Bagrus *meridionalis*.

Many cichlids are primarily herbivores feeding on algae (e.g. *Petrochromis*) and plants (e.g. *Etroplus suratensis*). Small animals, particularly invertebrates, are only a minor part of their diet. Other cichlids are detritivores and eat all types of organic material; among these species are the tilapiines of the genera *Oreochromis, Sarotherodon*, and *Tilapia*.

Other cichlids are predatory and eat little or no plant matter. These include generalists that catch a variety of small animals, including other fishes and insect larvae (e.g. *Pterophyllum*), as well as variety of specialists. *Trematocranus* is a specialised snail-eater, while *Pungu maclareni* feeds on sponges. A number of cichlids feed on other fish, either entirely or in part. *Crenicichla* are stealth-predators that lunge from concealment at passing small fish, while *Rhamphochromis* are open water pursuit predators that chase down their prey. Paedophagous cichlids such as the Caprichromis species eat other species' eggs or young, in some cases ramming the heads of mouthbrooding species to force them to disgorge their young. Among the more unusual feeding strategies are those of *Corematodus*, *Docimodus evelynae*, *Plecodus*, *Perissodus*, and Genyochromis spp., which feed on scales and fins of other fishes, a behaviour known as lepidophagy, along with the death-mimicking behaviour of *Nimbochromis* and *Parachromis* species, which lay motionless, luring small fish to their side prior to ambush.

This variety of feeding styles has helped cichlids to inhabit similarly varied habitats. Its pharyngeal teeth (teeth in the throat) afford cichlids so many "niche" feeding strategies, because the jaws pick and hold food, while the pharyngeal teeth crush the prey.

Reproduction

Figure: A substrate brooding female managuense cichlid, *Parachromis managuense*, guards a clutch of eggs in the aquarium.

Cichlids have highly organised breeding activities.

Brood Care

All species show some form of parental care for both eggs and larvae, often nurturing free-swimming young until they are weeks or months old. Communal parental care, where multiple monogamous pairs care for a mixed school of young have also been observed in multiple cichlid species, including *Amphilophus citrinellus*, *Etroplus suratensis*, and *Tilapia rendalli*. Comparably, the fry of *Neolamprologus brichardi*, a species that commonly lives in large groups, are protected not only by the adults, but also by older juveniles from previous spawns.

Several cichlids, including discus (*Symphysodon* spp.), some *Amphilophus* species, *Etroplus* and *Uaru* species feed their young with a skinsecreation from mucous glands. Parental care falls into one of four categories: substrate or open brooders, secretive cave brooders (also known as guarding speleophils), and at least two types of mouthbrooders, ovophile mouthbrooders and larvophile mouthbrooders.

Open Brooding

Open or substrate brooding cichlids lay their eggs in the open, on rocks, leaves, or logs. Examples of open brooding cichlids include Pterophyllum, *Symphysodon* spp, and *Anomalochromis thomasi*. Male and female parents usually engage in differing brooding roles. Most commonly, the male patrols the pair's territory and repels intruders, while females fan water over the eggs, removing the infertile and leading the fry while foraging. However, both sexes are able to perform the full range of parenting behaviours.

Cave Brooding

Secretive cave spawning cichlids lay their eggs in caves, crevices, holes, or discarded mollusc shells, frequently attaching the eggs to the roof of the chamber. Examples include Pelvicachromis spp., *Archocentrus* spp, and *Apistogramma* spp. Free-swimming fry and parents communicate in captivity and in the wild. Frequently this communication is based on body movements, such as shaking and pelvic fin flicking. In addition, open and cave brooding parents assist in finding food resources for their fry. Multiple neotropical cichlid species perform leaf-turning and fin-digging behaviours.

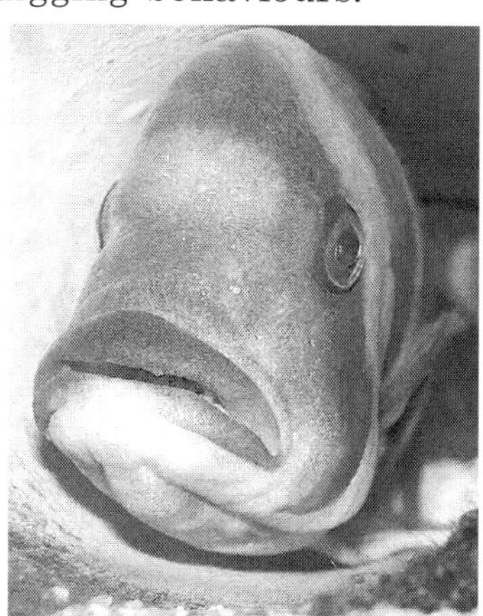

Figure: A female *Cyphotilapia frontosa mouthbrooding* fry, which can be seen looking out her mouth

Ovophile Mouthbrooding

Ovophile mouthbrooders incubate their eggs in their mouths as soon as they are laid, and frequently mouthbrood free-swimming fry for several weeks. Examples include many Great Rift Valley lakes (Lake Malawi, Lake Tanganyika and Lake Victoria) endemics, e.g.: *Maylandia*, *Pseudotropheus*, and *Tropheus*, along with some South American cichlids such as *Geophagus steindachneri*.

Larvophile Mouthbrooding

Larvophile mouthbrooders lay eggs in the open or in a cave and take the hatched larvae into the mouth. Examples include some variants of geophagus *altifrons*, and some *Aequidens*, *Gymnogeophagus*, and *Satanoperca*. Mouthbrooders, whether of eggs or larvae, are predominantly females. Exceptions that also involve the males include eretmodine cichlids (genera *Spathodus*, *Eretmodus*, and *Tanganicodus*), some *Sarotherodon* species, *Chromidotilapia guentheri*, and some *Aequidens* species. Rare paternal mouthbrooding occurs, for example, in *Sarotherodon melanotheron*. This method appears to have evolved independently in several groups of African cichlids.

Mating

Cichlids mate either monogamously or polygamously. The mating system of a given cichlid species is not consistently associated with its brooding system. For example, although most monogamous cichlids are not mouthbrooders, *Chromidotilapia*, *Gymnogeophagus*, *Spathodus* and*Tanganicodus* are all monogamous mouthbrooders. In contrast, numerous open or cave spawning cichlids are polygamous; examples include Apistogramma, *Lamprologus*, *Nannacara* and *Pelvicachromis*.

Population Status

In 2010, the International Union for Conservation of Nature classified 184 species as vulnerable, 52 as endangered, and 106 as critically endangered. At present, the IUCN only lists *Yssichromis sp. nov. "argens"* as extinct in the wild, and six species are listed as entirely extinct, but it is acknowledged that many more possibly belong in these categories (for example, *Haplochromis aelocephalus*, *H. apogonoides*, *H. dentex*, *H. dichrourus* and numerous other members of the genus *Haplochromis* have not been seen since the 1980s, but are maintained as Critically Endangered in the small chance that tiny –but currently unknown– populations survive).

Lake Victoria

Because of the introduced Nile perch (*Lates niloticus*) and water hyacinth, deforestation that led to water siltation, and overfishing, many Lake Victoria species have been wiped out or drastically reduced.

Figure: *Haplochromis latifasciatus* is critically endangered.

About two-thirds of endemic cichlids (approximately 300 species), especially bottom feeders, became endangered or extinct. Some survivors have adapted by becoming smaller or hybridising with other species. Satellite lakes such as Lake Edward and Lake Kyogahave not been as strongly affected, however, and harbour an array of similar species.

Food and Game Fish

Although cichlids are mostly small- to medium-sized, many are notable as food and game fishes. With few thick rib bones and tasty flesh, artisan fishing is not uncommon in Central America and South America, as well as areas surrounding the African rift lakes.

Tilapia

The most important food cichlids, however, are the tilapiines of North Africa. Fast growing, tolerant of stocking density, and adaptable, tilapiine species have been introduced and farmed extensively in many parts of Asia and are increasingly common aquaculture targets elsewhere.

Farmed tilapia production is about 1,500,000 tonnes (1,500,000 long tons; 1,700,000 short tons) annually with an estimated value of US$1.8 billion, about equal to that of salmon andtrout.

Unlike those carnivorous fish, tilapia can feed on algae or any plant-based food. This reduces the cost of tilapia farming, reduces fishing pressure on prey species, avoids concentrating toxins that accumulate at higher levels of the food chain and makes tilapia the preferred "aquatic chickens" of the trade.

Game Fish

Many large cichlids make good game fish. The strong, hard-fighting peacock bass (*Cichla* species) of South America is one of the most popular sportfish. It was introduced in many waters around the world. In Florida, this fish generates millions of hours of fishing and sportfishing revenue of more than US$8 million a year. Other cichlids preferred by anglers include the Oscar, Mayan cichlid (*Cichlasoma urophthalmus*), and jaguar guapote (*Parachromis managuensis*).

Aquarium Fish

Figure: The discus, *Symphysodon* spp., has been popular among aquarium enthusiasts.

Since 1945, cichlids have become increasingly popular as aquarium fish. Many cichlids are small to medium-sized, easy to feed with a range of prepared fish foods, breed readily, and practice brood care, making good aquarium fish.

The most common species in hobbyist aquaria is *Pterophyllum scalare* from the Amazon River basin in tropical South America, known in the trade as the "angelfish". Other popular or readily available species include the oscar (*Astronotus ocellatus*), convict cichlid (*Archocentrus nigrofasciatus*) and discus (*Symphysodon* spp.).

Cichlids can be kept in aquaria with other fish; however, many cichlids eat smaller fish. Conversely, some cichlids, such as *Apistogramma* or *Julidochromis* spp., can be timid. In such cases the use of dither fish is recommended.

Hybrids and Selective Breeding

Figure: The "red Texas" cichlid is not a Texas cichlid (*Herichthys cyanoguttatus*) but an intergeneric hybrid of *Herichthys* and *Amphilophus parents*.

Some cichlids readily hybridise with related species, both in the wild and under artificial conditions. Other groups of fishes, such as European cyprinids, also hybridise. Unusually, cichlid hybrids have been put to extensive commercial use, in particular for aquaculture and aquaria. The hybrid red strain of tilapia, for example, is often preferred in aquaculture for its rapid growth. Tilapia hybridisation can produce all-male populations to control stock density or prevent reproduction in ponds.

Aquarium Hybrids

The most ubiquitous aquarium hybrid is perhaps the blood parrot cichlid which is a cross of several species, especially from genus *Amphilophus*. With a beak-shaped mouth, an abnormal spine, and an occasionally missing caudal fin (known as the "love heart" parrot cichlid), the fish is controversial among aquarists. Some have called blood parrot cichlids "the Frankenstein monster of the fish world." Another notable hybrid, the flowerhorn cichlid, was very popular in some parts of Asia from 2001 until late 2003, and is believed to bring good luck to its owner. The popularity of the flowerhorn cichlid declined in 2004. Owners released many specimens into the rivers and canals of Malaysia and Singapore where they threaten endemic communities.

Numerous cichlid species have been selectively bred to develop ornamental aquarium strains. The most intensive programs have involved angelfish and discus, and many mutations that affect both colouration and finnage are known. Other cichlids have been bred for albino, leucistic, andxanthistic pigment mutations, including oscars, convicts and *Pelvicachromis pulcher*. Both dominant and recessive pigment mutations have been observed. In convict cichlids, for example, a leucistic colouration is recessively inherited, while in *Oreochromis niloticus niloticus* red colouration is caused by a dominant inherited mutation.

Figure: A leucistic long-finned form of the oscar, *A. ocellatus*

This selective breeding may have unintended consequences. For example, hybrid strains of *Mikrogeophagus ramirezi* have health and fertility problems. Similarly, intentional inbreeding can cause physical abnormalities, such as the notched phenotype in angelfish.

Marine Larval Ecology

Marine larval ecology is the study of the factors influencing the dispersing larval stage exhibited by many marine invertebrates and fishes. Marine organisms with a larval stage usually release large numbers of larvae into the water column, where these larvae develop and grow for a certain period of time before metamorphosing into adults. Most marine larvae are capable of dispersing long distances from their release site, although determining their actual dispersal distance is a significant challenge due to their microscopic size and the lack of an appropriate larval tracking method. Understanding dispersal distance, however, is important for a variety of reasons, including fisheries management, effective marine reserve design, and control of invasive species.

Theories on the Evolution of a Biphasic Life History

Marine larval dispersal is one of the most important topics in marine ecology today. Most marine invertebrates and many fishes have evolved a life cycle involving a demersal adult and apelagic larval stage or pelagic eggs that have the capacity to be transported long distances. There are several theories behind why these organisms have evolved this biphasic life history:

- Larvae use a different food source than adults, which may decrease competition between life stages.
- Pelagic larvae have the potential to disperse long distances, colonise new territory, and move away from habitat that has become overcrowded or otherwise unsuitable.
- A long pelagic larval duration can help a species break its parasite cycles.
- Pelagic larvae avoid benthic predators.

Pelagic larval dispersal, however, is not without its risks. For example, while larvae do avoid benthic predators, they are exposed to a whole new suite of predators in the water column.

Larval Development

Marine larval development can be broadly classified into three categories: direct development, lecithotrophic, and planktotrophic. Direct developers are characterised by a larval stage that has very low dispersal potential and usually looks like the adult form of the animal. These larvae are also known as "crawl-away larvae," since numerous marine snails exhibit this type of development, and their larvae crawl away from the egg mass. Lecithotrophic larvae generally have greater dispersal potential than direct developers. Many fish species and some benthic invertebrates have lecithotrophic larvae, which are provided with a source of nutrition to use during their dispersal, usually a yolk sac. Though some lecithotrophic species are capable of feeding in the water column, many, such as tunicates, are not, and must settle before depleting their food source. Consequently, these species have short pelagic larval durations and do not disperse long distances. Planktotrophic species, on the other hand, generally have fairly long pelagic larval durations and feed while in the water column. Consequentially, they have the potential to disperse long distances. This ability to disperse is one of the key adaptations of benthic marine invertebrates. During their time in the water column, planktotrophic larvae feed on phytoplankton and small zooplankton, including other larvae. Planktotrophic development is the most common type of larval development, especially among benthic invertebrates. The relatively long time most planktotrophic larvae spend in the water column and their apparently low probability of successful recruitment led some early researchers to develop a "lottery hypothesis" that states animals release huge numbers of larvae to increase the chances that at least one will survive, and that larvae cannot influence their probability of success. This hypothesis, though, views larval survival and successful recruitment as chance events, and numerous studies on larval behaviour and ecology have shown this to be false. Though it has been generally disproved, the larval lottery hypothesis does represent an important understanding of the difficulties faced by larvae during their time in the water column, particularly because it recognises the low probability of larval survival.

All three types of marine larvae face two major problems: avoiding predation and finding an appropriate site to settle.

Predator Avoidance

One of the major difficulties faced by larvae is the threat of predation. Larvae are small and plentiful, so many animals take advantage of this food source. The situation is particularly dangerous for invertebrate larvae in estuaries; estuaries are nursery grounds for planktivorous fishes. Estuarine species' larvae have evolved strategies to cope with this threat, including methods such as direct defence and avoidance. Direct defence is usually only evident in species in which larval development takes place entirely within the estuary. Studies have shown that larvae that do not leave estuaries are larger than larvae that develop in the open ocean. Additionally, many estuarine larvae have large spines and other protective structures. These defences work because most planktivorous fishes are gape-limited predators—what they eat is determined by how wide they can open their mouths—so larger larvae are harder for them to ingest. Morgan showed that spines do indeed serve a protective function by cutting off spines of some estuarine crab larvae and monitoring differences in predation rates between despined and intact larvae. Despined larvae suffered significantly higher predation rates than intact larvae, and were preferentially chosen during feeding trials with both types of larvae present. Additionally, Morgan showed that large-spined estuarine larvae usually keep their lateral spines relaxed, but raise them when approached by a predator. Therefore, predator deterrence in estuarine larvae is not only morphological but also behavioural.

A second strategy to deal with estuarine predators is to avoid them on small or large spatial scales. Some larvae do this at a small scale by simply sinking when approached by a predator. However, a more common avoidance strategy is to become active at night and remain hidden during the day, since most fishes are visual predators and need light to hunt. This strategy is not only evident in estuaries, but is also the main predator-avoidance strategy in the open ocean, since the water column lacks topography and thus hiding places. Most pelagic larvae and other planktonic species undertake diel vertical migrations between deeper waters with less light and fewer predators during the day and shallow waters in the photic zone at night, where their microalgal food source lives. By retreating to areas of low light during the day, marine larvae (and other zooplankton) can significantly decrease their risk of predation. On a larger scale, most estuarine invertebrate larvae avoid predators by leaving the estuary and developing in the open ocean, which has fewer planktivorous fishes. The most common strategy for leaving an estuary is reverse tidal vertical migrations. In this strategy, larvae use the tidal cycle and estuarine flow regimes to aid their departure to the ocean, a process that is well-studied in many estuarine crab species.

The process of reverse tidal vertical migrations begins when female crabs release larvae on a nocturnal spring high tide in an attempt to limit predation by planktivorous fishes. As the tide begins to ebb, larvae swim to the surface waters and are carried away from their site of hatching towards the ocean. When the tide reaches its low and begins to flood, larvae swim towards the bottom of the estuary, where water moves more slowly due to the boundary layer. This prevents them from being sloshed back and forth within the estuary in the surface waters. When the tide again changes back to ebb, the larvae swim to the surface waters and resume their journey to the ocean. Depending on the length of the estuary and the speed of the currents, this process can take anywhere from one tidal cycle to several days.

Dispersal and Settlement

Probably the most widely accepted theory explaining the evolution of a larval stage is the need for long-distance dispersal ability. Sessile organisms such as barnacles and tunicates, as well as sedentary species like mussels and crabs, need some mechanism to move their young into new territory, since they cannot move long distances as adults. Many species have relatively long pelagic larval durations—the amount of time a larva is in the water column before it is competent to settle—on the order of weeks or months. During this time in the water, larvae feed and grow, and many species move through several stages of development. For example, most barnacles molt through six naupliar stages before molting to a cyprid, the stage at which they seek an appropriate settlement substrate. This allows the larvae to use different food resources than the adults and gives them time to disperse.

This strategy, however, involves a certain degree of risk. While some larvae have been shown to be able to delay their final metamorphosis for a few days or weeks, few if any species are able to delay metamorphosis indefinitely, and most species cannot delay it at all. If these larvae metamorphose too far from a suitable settlement site, they perish. Due to the imperative of finding a suitable settlement site within a certain timeframe, many invertebrate larvae have evolved complex behaviours and endogenous rhythms to ensure their successful and timely settlement, which will be explained below.

While many estuarine species exhibit swimming rhythms of reverse tidal vertical migration to aid in their transport away from their hatching site, the same species can exhibit tidal vertical migrations to reenter the estuary when they metamorphose and are competent to settle. This process is similar to the reverse tidal vertical migrations described in the section discussing predator avoidance above, but instead of swimming down on flood tide, settlers remain in the surface waters, allowing themselves to be transported into the estuary.

Another change that many larvae undergo after they reach their final pelagic stage is to become much more tactile, clinging to anything larger than themselves. For example, Shanks observed crab postlarvae in the lab and found that they would swim vigorously until they encountered a floating object. Postlarvae would then cling to the object for the duration of the experiment. Shanks hypothesized that by clinging to floating debris, crabs can be transported towards shore due to the oceanographic forces of internal waves, which carry floating debris shoreward regardless of the prevailing currents.

If they are able to successfully return to shore, settlers encounter a new suite of problems concerning their actual settlement and successful recruitment into the population. Space is a limiting factor for sessile invertebrates on rocky shores, and larvae might not find any open habitat. Additionally, settlers must be wary of adult filter feeders, which usually cover the rocks at settlement sites and eat particles the size of larvae. Settlers must also avoid becoming stranded out of water by waves, and must select a settlement site at the proper tidal height to prevent desiccation and avoid competition and predation. To overcome many of these difficulties, some species rely on chemical cues to assist them in selecting an appropriate settlement site. These cues are usually emitted by adult conspecifics, but some species cue on specific bacterial mats or other qualities of the substrate.

Self-recruitment

One of the most important unanswered questions in larval ecology concerns the degree of self-recruitment in populations. For most of the short history of the field of larval ecology, larvae were considered to be passive particles that were carried by ocean currents to locations far from their site of hatching. This led to the belief that all marine populations were demographically open, connected by long distance larval transport. Recent work, however, is starting to show that many populations may be self-recruiting, and that larvae and juveniles are capable of purposefully returning to their natal sites.

Researchers take a variety of approaches to estimating population connectivity and self-recruitment, and several studies have demonstrated their feasibility. Jones et al. and Swearer et al., for example, both investigated the proportion of reef fish larvae returning to their natal reef after their time in the water column. Each study found higher than expected (possibly as high as 60%) self-recruitment in these populations, using variations of a typical mark, release, recapture sampling design. These studies were the first to provide conclusive evidence of self-recruitment in a species with the potential to disperse far from its natal site, and laid the groundwork for numerous future studies.

Implications

The principles of marine larval ecology can be applied to a number of fields both inside and outside the marine realm. Successful fisheries management relies heavily on understanding population connectivity and dispersal distances, and these processes are driven by larvae. Dispersal and connectivity must also be considered when designing natural reserves, both on land and in the water; if populations are not self-recruiting, then solitary reserves may lose their species assemblages. Additionally, many invasive species are able to disperse long distances during an early life stage, such as seeds in land plants or larvae in marine invasives. Understanding the factors influencing their dispersal is key to controlling their spread and managing already established populations. Through the continued study of the ecology of these microscopic creatures, scientists can better understand and more effectively manage myriad populations of both land and sea.

Extinct Fish

Extinct Jawless Fish

The first recorded jawless fish, known collectively as ostracoderms ýÿ i.e. shell-skinned, lived between 500 ýÿ 360 million years ago (mya). Ostracoderms had a bony armour, an internal cartilaginous skeleton and a heterocercal tail. Most lacked paired fins. The first recognisable lamprey was recorded in the Carboniferous period (325-380 mya). Although similar to modern lampreys, the relation of ostracoderms to living lampreys is unclear.

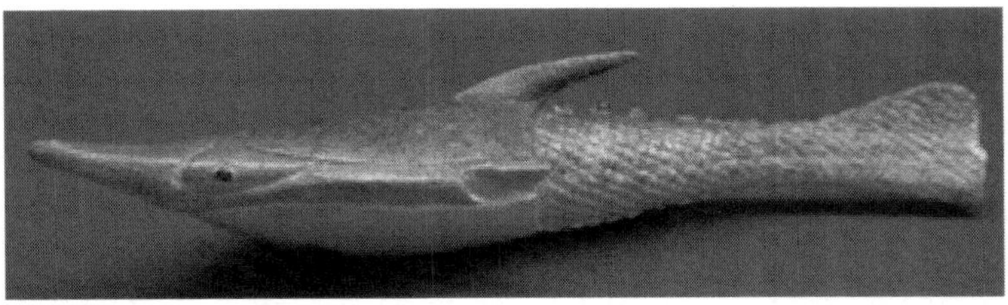

Figure: Reconstruction of *Pteraspis* sp. shows a finless ostracoderm, that lived in the Lower Devonian period (360 - 408 mya).

Figure: *Cephalaspids* were unusual ostracoderms in possessing pectoral fins. A fossil shows they were essentially nothing more than a large food-straining gill chamber with a swimming tail, nerves and sense organs. They lived from the Lower Devonian to the late Silurian (360 – 438 mya).

Extinct Jawed Fish: Acanthodians and Placoderms

The first group of jawed fish to appear in the fossil record around 440 mya were the acanthodians. Commonly called ýÿ spiny sharksýÿ, fish such as *Climatius* were small, large eyed, scaled fish. They had stout ornamented spines in front of all their fins and two rows of paired ventral fins. Acanthodians were plankton feeders living in marine and fresh-water habitats. Their relations to other fish is controversial. Certain features suggest they were an offshoot of bony fish; other features suggest that they were more closely related to cartilaginous fish. The second group of jawed fish, the Placoderms (plate-skinned), lived between 360 ýÿ 440 mya. They were bottom-dwellers with jaws that developed from the first gill arch and enabled them to prey on other animals. Placoderms appeared at the time when ancestors of modern groups would be expected to occur but show little evidence of being these ancestors.

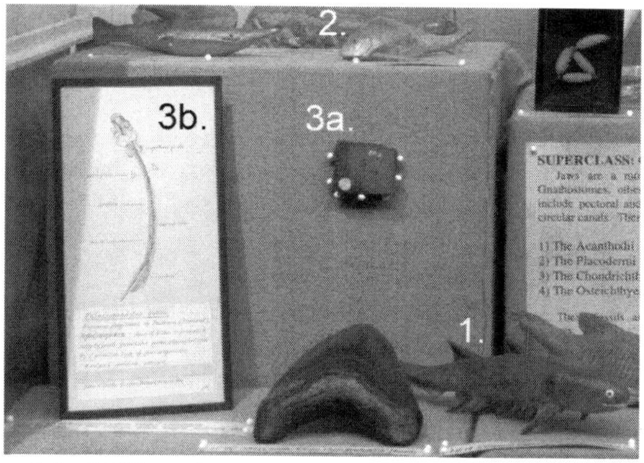

Figure: 1. A model of *Climatius* - an acanthodian - in life only 3 inches (8 cm) long. 2. Reconstruction of *Pterichthyes* - an antiarch. In this group of placoderms, the anterior end was enclosed in box plates. These fish are presumed to have crawled around on mud and rocks using their fins which were also enclosed in plates. 3. *Palaeospondylus gunni* is a tiny fossil animal from a Middle Devonian (360 - 408 mya) quarry in Scotland. Its identity is unclear: some suggest it is a larval form of a contemporary fish; others that it is a jawless fish or perhaps a placoderm without armour. An actual fossil shows the tiny size (1.5 cm); the drawing by M. Ramult shows the structure.

Bony Fish and the Ancestors of the Tetrapods

The different lineages of jawed bony fish that were ancestors of the modern teleosts, of the relict fish that exist today and of the tetrapods appeared in the Devonian period.

Near Extinct Fish Faces Death by Dynamite

Dynamited to death in the Mahaweli Ganga at Waratenna, Getambe, has been the fate of an 'endangered' endemic freshwater fish which has not been sighted for at least five years and was believed to be extinct in Sri Lanka.

While for three weeks blasts have been rocking the peaceful environs of Waratenna close to the Seemamalakaya on the Peradeniya-Katugastota Road, the sudden activity in connection with the groundwork for a mini-hydro power project has left the gadeya or Green Labeo (Labeo fisheri) dead, the Sunday Times learns.

The belief was that the gadeya was extinct worldwide, says researcher Pradeep Samarawickrama, explaining that about four months ago it was re-discovered by him at Waratenna. Earlier this fish was found in the Mahaweli system, at Polgolla and near Victoria, but with the construction of the dams under the accelerated Mahaweli Scheme, their habitats were submerged and changed.

Then it was assumed that the fish may be extinct but with Mr. Samarawickrama picking up the dead and decomposed fish it now indicates that the gadeya had made the Waratenna area its home.

"These sites should be preserved," he stressed, without indiscriminate dynamiting of the rocks embedded in the riverbed.

Another serious concern that this researcher raised was the impact of such blasting in an area which is heavily landslide-prone, with the government on one hand requesting people to be cautious.

Adding his voice to the concerns expressed by Mr. Samarawickrama including the need to preserve the 'last habitats' of this nearly-extinct fish, environmental lawyer Jagath Gunawardena quoting Section 27 of the Fisheries and Aquatic Resources Act said that any act or negligence that causes the death of any fish by the use of explosives is an offence. He questioned the integrity and inaction of the Mahaweli Authority, which being the project-approving body, has the responsibility of monitoring the implementation of the project that it had approved in the first instance.

Referring to the gadeya, Mr. Gunawardena said it was earlier found in areas such as Lewella, Tennekumbura and the lower regions of the Hulu Ganga but in the early 1980s it was predicted that 75% of the known habitats of this fish would go under water due to the construction of the Victoria reservoir.

"But now we realise the gadeya has been able to re-establish itself upstream," he said.

He lamented that due to short-sighted policies, these areas too would be destroyed. "L. fisheri has been assessed as 'endangered' due to its restricted range (extent of occurrence less than 5,000 km2), continuing decline in habitat quality and the fact that the species is considered to be found in less than five locations," according to the 2011 Global Red List of the International Union for Conservation of Nature (IUCN).

This species is found only in the mid to upper reaches of the Mahaweli Ganga basin, it states, explaining that L. fisheri may already be extinct, as a result of habitat loss following the Mahaweli Ganga development project.

Referring to habitat and ecology, the Red List states: "L fisheri is a benthopelagic species that only inhabits deep, fast-flowing sections of the river. It has not been recorded from unshaded, silty or turbid waters. It moves fast through the water, staying close to the bottom."

It was way back in 1980 that Ranil Senanayake predicted that the Polgolla and Victoria projects of the Mahaweli Scheme would destroy the gadeya's habitat while in 1990, Dr. Eric Wikramanayake said that this fish may already be extinct.

Bigger than Originally Believed

It was believed to grow only to 30 cm. but researcher Pradeep Samarawickrama has found a gadeya which is nearly twice that size, close to 60 cm. in length.

Sadly the fish is dead, due to companies which think only of short term gain and government authorities which do not do their jobs, said a conservationist while many others joined their voices to his, in pointing an accusing finger at the Mahaweli Authority for not acting quickly, even now, to stop the destruction of the gadeyas' habitat and re-evaluate the project.

Earlier, the known maximum length of this fish as reported by Dr. Paul Deraniyagala in his 'Vertebrates of Ceylon' Volume I published in the 1950s; the 'Freshwater Fishes of Sri Lanka' published by Rohan Pethiyagoda in 1991; and all other references was 30 cm.

However, a dead fish picked up by Mr. Samarawickrama is nearly twice that length.

Blasting Going on Despite Warning

The company involved in the project has been warned to halt the blasting with immediate effect, said Mahaweli Authority Director-

General D.M.C. Dissanayake who has sent a three-member inquiry team to the spot.

I am waiting the report, he said. However, the Sunday Times learns from residents in the area that work has not been halted and the blasting was going on even yesterday, with the whine of backhoes dumping the blasted rock onto one side continuing throughout the day, aggravating the already disturbed habitat.

Meanwhile, Central Environmental Authority Chairman Charitha Herath said that the CEA has not given its concurrence for the latest work at that spot.

The CEA has asked all stakeholders including the company involved to come for an urgent meeting on Tuesday, he said, detailing the background to the Sunday Times.

In 2007, a proposal was put forward to set up a mini-hydro power plant there and environmental approval sought. The Mahaweli Authority as the project-approving agency set the terms of reference for an environmental impact assessment (EIA). The EIA was reviewed by the Mahaweli Authority and the CEA asked for concurrence and as there were no public comments it was given, says Mr. Herath.

The concurrence was for a three-year period from February 2007 to February 2010, during which time they were supposed to build the plant. However, the project was not started and the approval expired on February 2010. Recently, CEA concurrence was requested once again, but we have not given it. That's why we called a meeting on July 26 to discuss it. However, we have received complaints that the project has been started.

Environmental lawyer Jagath Gunawardena, however, cites Section 24 B of the National Environment Act, according to which the CEA can issue a directive ordering the suspension of all activities of a project that is implemented without its permission. If this directive is not heeded the CEA can go to the Magistrate's Court and get it implemented as a court order, he adds.

An Emerging Role for a Cardiac Peptide Hormone in Fish Osmoregulation

It is clear from the extant literature that various fish groups face chronic osmoregulatory problems that depend on the surrounding salinity. Their physiologic and hormonal responses are largely those seen in the mammals, but their terrestrial descendants have lost osmoregulatory structures such as gills and rectal glands and depend primarily on renal function. A data base is now emerging that strongly suggests that a putative atriopeptin plays a role in osmoregulation in fishes. This conclusion is supported by the fact that heterologous AP produces relevant physiologic responses (e.g. natriuresis, vasodilation, stimulation, or inhibition of Na+ secretion by intestine, gills, and rectal gland) in both teleosts and elasmobranchs. Moreover, cardiac and brain extracts from fish can produce similar effects in both fishes and mammals, and these tissues from various fish groups contain immunoreactive AP, as does plasma.

Both physiologic and immunologic evidence suggests that the ventricle may be a significant source of AP in fishes, contrary to the situation in mammals. Finally, osmotic perturbations result in a change in plasma and tissue Apir levels. The finding that plasma APir levels increase in sea water, and that heterologous AP stimulates salt secretion by the teleost gill and shark rectal gland, and inhibits salt uptake by the teleost intestine, suggests that AP may primarily play a role in salt, rather than fluid, secretion in fishes.

The fact that in mammals AP inhibits prolactin secretion, but is itself stimulated by cortisol, supports this conclusion, since prolactin is generally considered to be the dominant osmoregulatory hormone in freshwater fishes, and cortisol serves this function in marine fishes. In addition, if AP inhibits brain AVT release in fishes, as it apparently inhibits vasopressin release in mammals, this also would be adaptive in marine fishes since AVT in fishes is diuretic, rather than antidiuretic. Interactions between AP and these hormones (prolactin, cortisol, and ATV) have not been studied in fishes to date, but these theoretical interactions do lend support to the hypothesis that AP may function primarily in salt homeostasis in fishes.

At least one potential hormonal interaction counters this argument, however. Atriopeptin is known to inhibit the production and effects of angiotensin II in mammals, and since this hormone is apparently dipsogenic in fishes, it may play a critical role in osmoregulation in sea water. Finally, it is of some historical interest that in Keys' original description of the eel heart-gill perfusion system in 1931 he commented that gill resistance remained constant for hours only if the heart itself was perfused

Membrane-Initiated Effects of Estrogen in the Central Nervous System

Hormonal ligands for the nuclear receptor superfamily have at least two interacting mechanisms of action: 1) classical transcriptional regulation of target genes (genomic mechanisms); and 2) nongenomic actions that are initiated at the cell membrane, which could impact transcription. Although transcriptional mechanisms are increasingly well understood, membrane-initiated actions of these ligands are incompletely understood. Historically, this has led to a considerable divergence of thought in the molecular endocrine field.

We have attempted to uncover principles of hormone action that are relevant to membrane-initiated actions of estrogens. There is evidence that the membrane-limited actions of hormones, particularly estrogens, involve the rapid activation of kinases and the release of calcium. Membrane actions of estrogens, which activate these rapid signalling cascades, can also potentiate nuclear transcription.

These signalling cascades may occur in parallel or in series but subsequently converge at the level of modification of transcriptionally relevant molecules such as nuclear receptors and/or coactivators. In addition, other hormones or neurotransmitters may also activate cascades to crosstalk with estrogen receptor-mediated transcription. The idea of synergistic coupling between membrane-initiated and genomic actions of hormones fundamentally revises the paradigms of cell signalling in neuroendocrinology.

Estrogens are Important endocrine effectors of reproduction. They also play an important and sometimes permissive role in liver and cardiovascular physiology, neuronal growth and differentiation, neuroprotection, cognition, and regulation of mood. They are ligands for estrogen receptors (ERs) α and β, members of the nuclear receptor superfamily. These isoforms, have a modular "domain" structure, a characteristic that they share with other members of the nuclear receptor superfamily.

The molecular mechanisms by which estrogens change the response of target tissues can be broadly divided into two categories: 1) genomic actions, and 2) nongenomic actions. The focus of this review is the interplay between both of these mechanisms with respect to transcription, particularly in the central nervous system (CNS). Each of the principles of estrogen action, stated as a section heading, is elaborated upon in that section.

Estrogens Signal via Both Genomic and Nongenomic Modes

The Genomic mode of Estrogen Action: Genomic actions are those wherein the ER exerts transcriptional effects via interaction with estrogens. In the absence of estrogens, the ER in various target tissues is associated with heat-shock proteins in a transcriptionally inactive state. Binding of 17β-estradiol, the natural endogenous ligand, or a similar agonist molecule induces a conformational change in the ER, promoting ER homodimerisation and subsequent nuclear translocation.

In the nucleus, the ER acts as a ligand-dependent transcription factor, binding estrogen response elements (EREs) regulating gene transcription. This mechanism is of importance in pathological states such as hormone-dependent breast cancers. Therapy that prevents hormone binding to ERα can decrease cell proliferation, whereas the presence of 17β-estradiol can increase the proliferation rate.

In the CNS, the focus of this review, several genes are also regulated in this manner by estrogen-bound ERα. For example, in ERα knockout mice, estrogen cannot induce the progesterone receptor (PR) gene in the hippocampus, indicating that genomic effects on this important estrogen-marker gene required the ERα. Increased transcription of neurotrophins such as brain-derived neurotrophic factor (BDNF) and their high-affinity receptors, the oxytocin-oxytocin receptor genes, the preproenkephalin gene, the antiapoptotic genes such as Bcl-2, are important in neuroprotection, neurodevelopment, and reproductive behaviours driven by estrogens. Generally, cellular signalling cascades activated by estrogens and other hormones were thought not only to be disparate from classical genomic actions but also to lead to fewer or less important phenotypes.

Membrane-initiated (Nongenomic) Actions of Estrogens

First reported in 1967 by Szego and Davis, acute 17β-estradiol administration could increase cAMP in the uterus of ovariectomised mice within 15 sec, in stark contrast to the slower genomic actions of estrogens, which take hours for final changes in protein expression to occur, after transcriptional regulation. Nongenomic actions of estrogens can therefore be defined as rapid (milliseconds to minutes) effects, generally initiated at the plasma membrane by 17β-estradiol, resulting in the activation of signal transduction pathways, including calcium flux, within target cells. The nature of the receptor that binds 17β-estradiol at the plasma membrane is immaterial to this definition. Studies that attempt to conclusively demonstrate membrane-initiated estrogen effects that are confined to the membrane often utilise membrane-limited estrogen conjugates, such as E2-BSA (17β-estradiol linked to BSA) or E2-HRP (17β-estradiol linked to horseradish peroxidase) to separate membrane-initiated actions of these conjugates from effects that arise from the steroid inside the cell. These compounds consist of 17β-estradiol molecules (32–38 moles of 17β-estradiol/BSA conjugated to the BSA molecule at a specific site, such as to the sixth carbon via an oxime linkage, to the 17th carbon or to the third carbon. In the immortalised GT-1 cell line, both 17β-estradiol and E2–6-BSA increase cAMP. Similarly, in primary cultures of neostriatal neurons, both 17β-estradiol and E2–6-BSA reduce calcium currents.

Although most studies have used one or sometimes several of these compounds to mimic a membrane-limited action for 17β-estradiol, several caveats should be noted regarding the use of these conjugates. There has been considerable controversy over the possible release of free 17β-estradiol from membrane-limited conjugates and the possibility of endocytosis of the conjugate itself with the subsequent release of free 17β-estradiol into cells. Stevis *et al.* demonstrated that both E2–6-BSA and E2–17-BSA had free, presumably genomically active, 17β-estradiol in the preparation as measured by RIA and required filtration to remove the free 17β-estradiol.

This suggests that, in some studies, signal transduction pathways that were thought to have been initiated at the membrane might have been initiated within the cell by free 17β-estradiol present in the unfiltered E2-BSA preparation. In addition, the activity of E2-BSA conjugates has not always paralleled the activity of free 17β-estradiol itself. For example, E2-BSA devoid of free 17β-estradiol (achieved by filtration) did not bind either ERα or ERβ or to an ERE under *in vitro* conditions where 17β-estradiol did bind both ER isoforms and the ERE. In the SK-N-SH neuroblastoma cell line, filtered E2-BSA induced MAPK, whereas free 17β-estradiol did not do so. This suggests that the E2-BSA binding conformation to the ER differs from that of 17β-estradiol with a subsequent change in the ability of the receptor to bind or transactivate genes that are regulated by EREs. This could result in artifactual results, *i.e.*, nonphysiological activation of rapid signal transduction pathways, with E2-BSA if ERα or ERβ acts as the membrane ER.

Within neurons, studies from Susan Wray's laboratory also show functional differences between different E2-BSA conjugates in the activation of the protein kinase A (PKA) pathway and calcium flux. Although both 17β-estradiol and filtered E2–17-BSA increased intracellular Ca levels in GnRH-positive cells in nasal explants, the transcription inhibitor, 5,6-dichloro-1-β-D-ribobenzimidazole (DRB) could abrogate the 17β-estradiol-mediated increase but not the E2–17-BSA increase in calcium in explants. In a primary culture of GnRH neurons, E2–6-BSA did not increase cAMP response element (CRE) binding protein (CREB) phosphorylation although 17β-estradiol did so. This suggests that the mode of conjugation of the 17β-estradiol moiety to BSA may allow different conjugates to contact different membrane ERs, *i.e.*, different proteins. Alternatively, binding of free 17β-estradiol and E2-BSA conjugates to the same receptor protein could lead to differences in protein conformation, which presumably lead to differences in calcium flux.

All of these studies using filtered E2-BSA conjugates and 17β-estradiol demonstrate that different signal transduction pathways could be activated by different E2-BSA conjugates or 17β-estradiol, depending on cell type. One implication is that different cell types possess varied combinations of proteins capable of binding E2-BSA and/or free 17β-estradiol. Recently, the Katzenellenbogen laboratory has synthesized estrogen dendrimer conjugates that are stable, do not seem to leach free 17β-estradiol, and do not enter the nucleus of breast carcinoma cells to conduct genomic actions.

They do not induce transcription of endogenous genes and can activate the ERK pathway rapidly at 2 min. after treatment at levels similar to free 17β-estradiol. These studies underscore the importance of using free 17β-estradiol (the endogenous ER ligand) as a comparison to the artificial conjugate, E2-BSA in all studies. If the actions of the artificial conjugate, E2-BSA, and 17β-estradiol are comparable, the E2-BSA preparation should preferably have been filtered and demonstrated to possess no transcriptional activity to conclude a true nongenomic, membrane-initiated effect. Unfiltered E2-BSA could show effects similar to 17β-estradiol due to the presence of free 17β-estradiol in the preparation. It should be noted that very few studies, particularly those before 2000, have explicitly shown that the E2-BSA preparation used in the study has no free 17β-estradiol or no genomic activity.

What, therefore, could be considered a nongenomic effect? If filtered E2-BSA is not used in the study, an acute application of 17β-estradiol that activates classically nongenomic effects (*i.e.*, kinase activation or calcium flux) in a short time frame could also be considered to be a membrane-initiated action of estrogen. However, calcium flux could also be regulated nongenomically by 17β-estradiol acting directly on intracellular calcium stores, apart from being initiated at the membrane. Similarly, kinase activation could also be a result of stabilisation or localisation; these processes may be nongenomically activated by estrogens but may not be plasma membrane-initiated.

Membrane-initiated (Nongenomic) Actions of Estrogens in Neural Cells

In this review, we will discuss nongenomic actions of estrogens in the CNS only when they occur in physiologically relevant contexts, which will also be described to lend perspective to these actions. In the CNS, membrane-initiated actions of estrogens have been observed predominantly in the context of reproduction, neuroprotection, and neurotrophism, all of which are also affected significantly by the genomic actions of estrogens.

Nongenomic Actions Play a Dominant Role in the Estrogen Effect on Neuronal Excitability

Estrogens have a rapid effect on neuronal excitability; this has been typically studied using neurons from the hypothalamus relevant to reproductive physiology, such as those that produce GnRH or neurons in the ventromedial hypothalamus (VMH) that control reproductive behaviour or lordosis. In hypothalamic slice cultures from ovariectomised guinea pigs, 17β-estradiol could uncouple the ¼-opioid receptor in pro-opiomelanocortin (POMC) neurons and the ³-aminobutyric acid (GABA) (B) receptors from dopaminergic neurons from the small conductance, G protein-coupled inwardly rectifying K^+ channels (GIRK), resulting in lower efficacy of ¼-opioid and GABA agonists activation of these channels and lower hyperpolarisation of these cells. This increase in 17β-estradiol-mediated excitation is blocked by a variety of inhibitors to intracellular signalling cascades, such as phospholipase C, PKA, and protein kinase C (PKC) (46, 47).

The uncoupling effect can be mimicked by membrane impermeant E2-BSA and the selective ER modulators, raloxifene and 4-hydroxytamoxifen. The inhibitor data, along with the blockade of uncoupling by an ER antagonist, ICI 182,780, suggest that the ER is G protein-coupled in hypothalamic neurons. An alternate explanation could be that the ER is required intracellularly for this effect, whereas the rapid signalling cascades are activated independently by a membrane ER (mER) coupled to a G protein. Qiu *et al.* have proposed a model whereby a $G\alpha_q$ coupled ER at the membrane initiates this signalling cascade of enzymes, which appear to be arranged in series. Using subtractive hybridisation and microarray analysis, Malyala *et al.* showed that estrogen altered the mRNA levels for a number of neurotransmitter receptors and synaptic vesicle release molecules [*e.g.*, GABA (B) R2 subunit and synaptobrevin] as well as for enzymes in rapid signalling cascades [*i.e.*, phosphotidylinositol-3-kinase (PI3K) p55[3], gec-1, Rab 11a GTPase]. Hence, these molecules could be regulated at the level of enzyme activation by rapid, membrane-initiated mechanisms and at the level of transcription by genomic mechanisms. However, at least some part of the effect of 17β-estradiol on neuronal excitability is solely rapid and membrane-initiated, as evidenced by both the short time scales of application and the data that cycloheximide use in hypothalamic slices could not block the uncoupling of ¼-opioid receptors to the GIRK channels.

Contrary to the evidence that a $G\alpha_q$ receptor may increase excitability in the arcuate nucleus, in CA1 hippocampal neurons, this could be attributed to a $G\alpha_s$ receptor that is activated rapidly by estrogens. 17β-Estradiol as well as E2-BSA rapidly potentiated kainate-induced currents via PKA activation initiated by a $G\alpha_s$ receptor. This was preserved in both wild-type and ERα knockout mice, suggesting that the classical ER was not required for the increase in excitability in CA1 hippocampal cells in response to kainate. The increase in neuronal excitability by estrogens is a scenario where nongenomic effects of estrogens are probably dominant over genomic effects.

Effects of Estrogen on Neuroprotection Involve the Activation of Kinases, Calcium Flux, and the Induction of Genes

Estrogen-mediated neuroprotection against various agents such as ischemia, glutamate, and *N*-methyl-D-aspartate (NMDA)-mediated excitotoxicity and degenerative diseases such as Parkinson's and Alzheimer's disease has attracted intense interest. A majority of this work stems from the finding that intact female rats had lower mortality and better neuronal survival than male rats after middle cerebral artery occlusion (MCAO) but that this effect was eliminated by ovariectomy.

Administration of 17β-estradiol, but not progesterone, to ovariectomised female rats, aged reproductively senescent rats, and male rats reduced infarct volumes after MCAO. In general, to achieve neuroprotection against excitotoxicity and β-amyloid-mediated toxicity in neurons, estrogens regulate molecules that are involved in programmed cell death and inflammation in a multifaceted manner using both genomic and membrane-initiated modes of action. Central to neuroprotection in hippocampal neurons and in NT2 cells is the up-regulation by 17β-estradiol of the antiapoptotic genes Bcl-2 and

Bcl-XL and the down-regulation of Nip2, an inhibitor of Bcl-2, as well as the activation of PI3K and the downstream substrate, protein kinase B (Akt/PKB). In addition, 17β-estradiol induction of Bcl-2 mRNA has been shown to require IGF-I receptor activation; ERα is abundantly coexpressed with IGF-I receptor in female rat brain. Because both 17β-estradiol and ERα are needed, presumably ligand-dependent transcription by the ERα is needed.

However, the requirement for IGF-I receptor activation is less clear; it is possible that the downstream signalling events from the IGF-I receptor, *i.e.*, MAPK, and Akt kinase activation are needed for Bcl-2 induction. Indeed, in PC12 cells, estrogen-mediated Akt/PKB activation induces Bcl-2 mRNA via the phosphorylation of the CREB and the subsequent binding of this protein to the CRE in the Bcl-2 promoter. Recently, Wu *et al.* have reported that calcium influx mediated by rapid regulation of L-type calcium channels by 17β-estradiol is critical in initiating a src/ERK signalling cascade that also results in enhanced Bcl-2 expression in cultured rat hippocampal neurons. Mendez *et al.* have proposed a hypothetical model for IGF-ER interaction in the brain that is relevant to neuroprotection.

On activation of the IGF-IR by IGF, the ERα associates with the receptor at the plasma membrane and activates PI3K via interaction with the p85 subunit of PI3K. Regulation of the downstream Akt/PKB target, glycogen synthase 3β (GSK-3β), by 17β-estradiol results in a decrease in phosphorylated Tau protein in the rat hippocampus, providing protection against Alzheimer's disease. However, it is not clear whether the dynamics of this model is similar with ligand-free ERα or ligand-bound ERα or how the activation of GSK-3β decreases Tau protein.

Rapid activation of both PKB and MAPK, and to a lesser extent PKC, by estrogens has also been shown to be important in neuroprotection in a number of neuronal cell types and against different types of insults. In the rat retina, rapid activation of the PI3K pathway is important in estrogen-mediated protection against light-induced photoreceptor degeneration. 17β-Estradiol also rapidly activates p44/p42 MAPK in cortical slices and in hippocampal neurites. In the rat hippocampus, rapid induction of MAPK by estrogens protects against quinoline-induced toxicity, whereas in hippocampal primary culture as well as in SK-N-MC neuroblastoma cells, phosphorylation of the MAPK ERK 1/2 by estrogens leads to enhanced secretion of β-amyloid precursor protein (sAPP). In primary rat cortical neurons, application of 17β-estradiol or a membrane-limited E2-BSA conjugate rapidly stimulated the release of sAPPα, whereas PKC inhibitors, added in addition to estrogens, block this effect.

This secreted product decreases the generation of the β-amyloid protein itself, which is a dominant neurotoxic component in the plaques of Alzheimer's patients. Exogenous 17β-estradiol treatment did not confer neuroprotection against MCAO in PKC[3] knockout mice, suggesting that this PKC isoform mediates neuroprotection by 17β-estradiol against ischemia. Although these studies show that both 17β-estradiol-activated MAPK and Akt pathways appear to be necessary for neuroprotection, alone they are not sufficient. Although Bcl-2 transcription appears dependent on rapid signalling pathways, not all genes involved in neuroprotection may be so dependent, and not all signal transduction cascades may affect transcription. In addition, the relative contribution of membrane-initiated *vs.* genomic actions in neuroprotection is unclear.

Effects of Estrogen on Dendritic Spines Involve Both Nongenomic and Genomic Actions

17β-Estradiol also increases the growth of dendritic spines, modulating synaptic plasticity, in several hormone-responsive brain areas such as the VMH, hippocampus, and amygdala via both genomic and membrane-initiated mechanisms. Estrogen-mediated development of spines in rat hippocampal cultures is dependent on rapid NMDA receptor activation via c-src, which in turn may be activated by 17β-estradiol binding a putative membrane ER. Similar to the scenario in midbrain dopaminergic neurons, PKA activation and phospho-CREB (pCREB) increases are also required for 17β-estradiol-induced spine increase in rat hippocampal primary cultures.

Estrogen-induced pCREB in the rat hippocampus, which occurs as early as 1 h after addition of 17β-estradiol, requires calmodulin kinase (CaMK)II and MAPK signalling but is less dependent on PKB. However, the longer time frame of inhibitor addition in this study does not allow conclusive confirmation of a rapid or membrane-limited effect of estrogen on these pathways, although it is highly probable, based on the signalling pathways activated. In female mice, administration of 17β-estradiol for 1 h increased pCREB immunoreactivity in the VMH and medial preoptic area, whereas in male mice, pCREB was increased in the CA1 field of the hippocampus and in the medial septum. In cultured rat hippocampal neurons, 17β-estradiol administration increased MAPK-dependent CREB activation in as little as 3 min. In differentiated NG108–15 neuroblastoma cells, Akt/PKB activation by 17β-estradiol increased protein synthesis of PSD-95, a major component of spines, through phosphorylation of the elongation factor, 4E-BP1; this is clearly a nongenomic effect because PSD-95 mRNA does not appear to be regulated by estrogen in this system.

Estrogen-induced stabilisation and maintenance of spines may also depend on the decrease of GABA inhibition exerted by interneurons on CA1 pyramidal neurons via decrease in glutamate decarboxylase content in interneurons by 17β-estradiol. Such a GABA decrease could also increase BDNF synthesis by solely genomic mechanisms such as transcriptional up-regulation of BDNF mRNA via a noncanonical ERE in hippocampal pyramidal cells.

A more controversial model for 17β-estradiol-mediated increase in BDNF levels proposes that membrane ER signalling results in BDNF translation in giant mossy fibre boutons, which in turn synapse on CA3 pyramidal neurons. In differentiating midbrain dopaminergic neurons, 17β-estradiol rapidly (within 30 min.) activates the PI3K/Akt signalling cascade; both this pathway and estrogen-induced PKA are important in neurite outgrowth in these neurons.

Blockade of PKA or NMDA receptor or BDNF abrogates the effect of estrogens on spines in hippocampal neurons, suggesting that all of these pathways are needed for spine formation and stabilisation. Alternatively, these could be part of a single pathway that leads to spine formation; hence, decreasing the levels of one molecule in the pathway could decrease the number of spines dramatically.

Because all of these studies, with the exception of studies that involve pCREB and the PKA signalling pathway, use application of free 17β-estradiol on longer time scales than is traditional for membrane-initiated actions, it is impossible to conclusively rule out transcriptional genomic actions of 17β-estradiol on these molecules.

However, the requirement for the activation of several kinases suggests that nongenomic pathways are important in dendritic spine growth, in concert with genomic effects of estrogens.

The data in these studies suggest that the PI3K/PKB pathway may be predominant in neuroprotection, whereas the PKA pathway could be more important in neurotrophism. Both signal transduction pathways, initiated by G protein-coupled membrane receptors, affect neuronal excitability in the hypothalamus and hippocampus.

Membrane-Initiated Hormone Actions Potentiate their Transcriptional Actions

The hypothesis that rapid signalling cascades initiated at the membrane by estrogens can lead to changes in transcription from ERE-containing genes is inspired by the fact that reproductive behaviour or lordosis in female mammals is dependent on gene expression in discrete brain areas as well as kinase activation and changes in neuronal excitability by estrogens. Hence, this is a good model system to study the interaction between rapid and slow actions of estrogens.

Coupling of the Membrane-initiated Actions of Estrogen to Transcription: The Consequence for Reproductive Behaviour

The estrogen control of female reproductive behaviour or lordosis presents a scenario for the interplay between membrane-initiated signal transduction mechanisms and the classical genomic mode of transcription. In the VMH, estrogen imposes hormonal dependence on neurons that control descending efferents to the muscles in the lower back that allow for the arching of the back, a posture characteristic of lordosis. Genes transcriptionally regulated by estrogen that may play a role in approach and sex behaviours in rodents include the PR; the oxytocin receptor and its ligand, oxytocin; the preproenkephalin peptides; and the α1b-adrenergic and muscarinic receptors.

In support of the idea that membrane-initiated actions play a role in rodent lordosis behaviour, second messengers in kinase cascades can potentiate lordosis. For example, phorbol esters (PKC activators) infused into the midbrain central gray could facilitate lordosis, whereas PKC inhibitors could reduce lordosis behaviour. Dibutryl cAMP and 8-bromo cAMP infused into the VMH could reduce the inhibition mediated by serotonin, suggesting that the coupling of the serotonin receptor to PKA inhibition has physiological relevance.

Antagonists to protein kinases or to G protein-coupled receptors (GPCRs) reduce lordosis behaviour, suggesting that activation of kinase cascades is permissive to lordosis. Infusion of PI3K or MAPK inhibitors into the VMH during estrogen priming of ovariectomised rats attenuates lordosis. In addition, IGF receptor antagonists block lordosis induction by estrogen in female rats, whereas δ-opioid receptor agonists infused into the VMH, but not into the medial preoptic area, can increase lordosis in female rats already primed with estrogen and progesterone. In the VMH, estrogen increase in neuronal excitability is thought to be relevant to lordosis behaviour in rodents. Similar to the effects of estrogen on POMC and dopamine neurons, in VMH slices, acute 17β-estradiol administration (10 min.) can potentiate the neuronal excitation caused by both histamine and NMDA, two neurotransmitters that act via GPCRs and ligand-gated ion channels, respectively.

This suggests that 17β-estradiol can influence two different transduction pathways to achieve a final effect, *i.e.*, increase in depolarisation of neurons. Because it appears that both genomic and membrane-initiated actions of estrogen are possibly important in lordosis behaviour, we initially hypothesized that membrane-initiated actions of estrogen impacting signal transduction cascades could regulate gene transcription, also induced by estrogen.

Coupling of Membrane-initiated Actions of Estrogens to Transcription in Nonreproductive Functions

Apart from lordosis behaviour, the above hypothesis also has roots in neuronal and non-neuronal cell culture studies which show that estrogen-induced gene transcription is indeed influenced by signalling molecules. In the human neuroblastoma cell line, SK-N-SH, a membrane-limited estrogen conjugate (E2-BSA) can induce the c-fos gene via MAPK kinase and MAPK, demonstrating a mechanism by which genes possessing non-ERE promoters might still be induced by estrogen. In the same cell line, E2-BSA could induce the neurotensin gene, an effect blocked by PKA inhibitors.

E2-BSA alone did not induce transcription from the consensus ERE; presumably, there was no free 17β-estradiol in this preparation. In the MCF-7 breast carcinoma cell line, PR mRNA is induced by both estrogen and PKA activators, whereas the estrogen up-regulation of cyclin D1 has been reported to be through the CRE. The levels of estrogen-regulated mitogenic proteins, cyclin D1 and c-myc, can be decreased by inhibition of MAPK and PI3K.

Inhibitors to PI3K, dominant-negative Akt mutants, and a selective ErbB2 inhibitor can decrease the up-regulation of ERα mRNA by estrogen (autoregulation) in MCF-7 cells. In endothelial cells, inhibitors for PI3K or the use of dominant-negative constructs to PI3K also block estrogen-induced cyclo-oxygenase-2. An inhibitor of PKC could decrease transcriptional potentiation by E2-BSA of ERα-mediated transcription from a consensus ERE in MCF-7 cells. These studies show that a number of estrogen-induced genes or estrogen-induced promoters can be regulated by second messenger pathways.

The Coupling of Membrane-initiated Actions to Transcription is Studied in Vitro by Using a Two-pulse Transcriptional Paradigm in Cell Culture

A "proof of concept" result for the hypothesis that transcription can be influenced by kinase cascades was first demonstrated in a neuroblastoma cell line, SK-N-BE(2)C, which is devoid of endogenous ERα or ERβ isoforms and has been used previously as a model for estrogen action in the brain upon the transient or stable transfection with ERα. This cell line has been studied extensively for the effects of both ¼- and δ-opioid receptors, demonstrating that GPCR-mediated receptor effects are possible in this cell line.

The use of a novel transcriptional two-pulse hormonal administration paradigm in cell culture as a tool to study transcriptional potentiation was inspired by earlier studies where discontinuous, pulsatile estrogen administration is effective in achieving cellular division and behaviour. In a pioneering study, Harris and Gorski showed that a discontinuous schedule of estrogen administration can induce uterine cell division. Discontinuous schedules of two 1 h pulses of estradiol benzoate (EB) with an interpulse interval of 6 h are as effective in inducing both lordosis and the PR in the VMH of female rats 48 h after the first EB administration as a continuous administration of EB for 48 h. It is possible that in the first pulse, estrogen may initiate kinase cascades (rapid membrane-initiated actions) that may potentiate genomic transcriptional actions in the second pulse.

To separate the membrane-initiated actions of estrogen from the genomic slower actions of estrogen in this cell line, estrogens added in the first 20 min. pulse can initiate signalling from the membrane, whereas 17β-estradiol in the second 2 h pulse promotes genomic action on a consensus ERE-linked reporter in neuroblastoma cells. We cannot completely rule out membrane actions in the second pulse because free 17β-estradiol also has the ability to act at membrane sites. To initiate signalling cascades solely at the membrane, a membrane-impermeable estrogen conjugate, E2-BSA, filtered to remove free 17β-estradiol, was applied in the first pulse. Despite controversy over the possible leaching of free 17β-estradiol into cells, we have shown that continuous administration of E2-BSA for 24 h to transfected SK-N-BE(2)C cells does not increase transcription above basal levels from a consensus ERE-based reporter gene. In addition, Taguchi *et al.* showed that E2-BSA possessed the ability to bind to the ER at membrane sites.

In these initial experiments, nanomolar concentrations of E2-BSA can potentiate transcription from a transfected consensus ERE-driven luciferase gene in neuroblastoma cells induced by a second 2 h pulse of 10^{-9} M 17β-estradiol (E2-BSA-mediated transcriptional potentiation). E2-BSA can hence substitute for 17β-estradiol in the first pulse.

A lower level of transcription (albeit higher than control) is seen with vehicle in the first pulse, followed by the addition of 17β-estradiol in the second longer pulse. A "reverse" pulse paradigm where 17β-estradiol is given in the first pulse followed by E2-BSA in the second pulse does not potentiate transcription, demonstrating that preliminary membrane effects are both sufficient and essential for later transcriptional effects.

To pinpoint specific kinases that play a role in E2-BSA transcriptional potentiation, specific inhibitors in the first pulse along with E2-BSA or in the second pulse with 17β-estradiol were used. This strategy demonstrated that PKA and PKC may play a role in E2-BSA-mediated transcriptional potentiation because the PKC and PKA inhibitors chelerythrine and Sp-cAMPs in the first pulse reduce the level of transcription but do not interfere with potentiation when added in the second pulse.

On the other hand, the calcium chelator BAPTA-AM blocks E2-BSA-mediated transcriptional potentiation when added either in the first pulse along with E2-BSA or in the second pulse with 17β-estradiol, suggesting that calcium is important in both the membrane-initiated and genomic actions of estrogen.

Using selective inhibitors, we have also provided evidence that the entry of calcium via N-type voltage-gated calcium channels is important in both genomic and membrane-initiated actions of estrogens, whereas the ryanodine receptor, which modulates intracellular stores of calcium, is important in coupling membrane-initiated actions of estrogen initiated at the membrane to transcription.

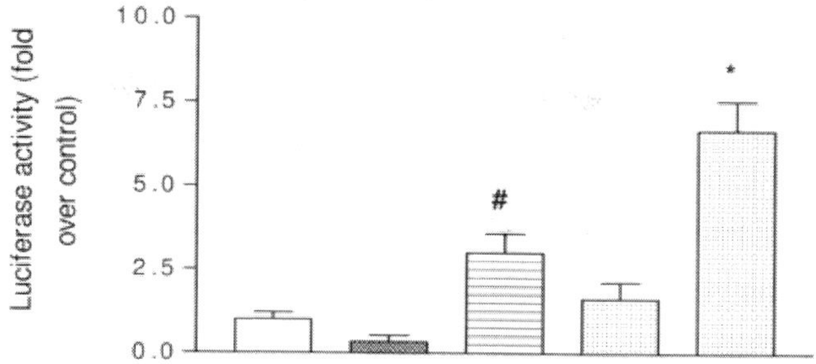

1st pulse	E-BSA	-	+	-	-	+
	E_2-10^{-9}M	-	-	-	+	-
2nd pulse	E-BSA	-	-	-	+	-
	E_2-10^{-9}M	-	-	+	-	+

Figure: E2-BSA (10^{-9} M) potentiates transcription mediated by 17â-estradiol (10^{-9} M) in the second pulse. SK-N-BE(2) C neuroblastoma cells are cotransfected using Effectene reagent (QIAGEN Inc., Valencia, CA) with pGL2-TATA-Inr-Luc (reporter; 3xereluc) (200 ng), pSG-hERá (80 ng), pSV-âgal (80 ng) (Promega, Madison, WI), and pBSKII+ (Stratagene, La Jolla, CA) to a total of 400 ng/well. Twenty-four hours after transfection, a two-pulse regimen is initiated with both pulses separated by a hormone-free 4 h interval. Results (n = at least 4 per treatment group in duplicate experiments) are analysed using one-way ANOVA followed by Student-Newman-Keuls *post hoc* test (Prism Software Corp., Irvine, CA) to compare normalised luciferase activity between treatment groups. E2-BSA (10^{-9} M) was given for 20 min. in the first pulse, and 17â-estradiol (10^{-9} M) was given in the second pulse for a duration of 2 h. Test of "potentiation" hypothesis: E2-BSA given in the first pulse was followed by 17â-estradiol in the second pulse (*far right bar*). Results (n = 4 per treatment group from replicate experiments) are represented as mean + SEM. *, $P < 0.05$ compared with all other treatment groups; #, $P < 0.05$ compared with the vehicle group.

The Signal Transduction Cascades Initiated by Estrogens at the Membrane May Occur in Parallel or in Series

The experimental strategy using inhibitors to different kinase cascades and calcium channels resulted in several kinases being implicated in E2-BSA-mediated transcriptional potentiation. Presumably, all these kinases are activated by estrogens acting at the membrane of neuroblastoma cells. In addition, our data (unpublished) with dominant-negative and constitutively active Gα mutants demonstrate that the membrane ER in this cell line may be coupled to Gα_q. Attempting to present signalling hierarchies in this cell line, models were constructed taking into account both the signalling cascades elucidated by us and previous work from Martin Kelly's laboratory which suggests that PKC activation of PKA via adenylyl cyclase (signalling in series) underlies the modulation of the SK channel in the guinea pig arcuate nucleus by estrogen. Qiu *et al.* have also suggested that estrogen can modulate ¼-opioid and GABA activity in the arcuate nucleus via an ER coupled to $G\alpha_q$. What is the role for calcium in our paradigm?

The absolute importance of calcium is shown by the use of BAPTA-AM; this calcium chelator abrogates all transcription when added in either the first pulse with E2-BSA or in the second pulse with 17β-estradiol. In the mouse hippocampus, administration of EB for 30 min. activates CaMKII in a time- and dose-dependent manner, whereas CamKIV is an estrogen target in the rat medial amygdala. However, our data (unpublished) using inhibitors indicate that CaMKII is not important in E2-BSA-mediated potentiation of transcription in neuroblastoma cells.

PKC activation by calcium is yet another possibility. In addition, calcium is important in the stabilisation of the ERα in MCF-7 human breast cells and may be a point of convergence for all the kinase cascades. The ability of all other single kinase inhibitors to reduce, but not totally negate, the E2-BSA-mediated transcriptional potentiation suggests that transcription is a sum of signalling from several kinase cascades. This implies that signal transduction pathways may originate in parallel from a single signal at the membrane. However, it is also possible that some of these kinases could be needed to maintain the cell in a basal responsive state or be permissive, but not essential, to the membrane-initiated actions of estrogens.

Membrane-initiated Actions of Estrogens may also Occur in Parallel or in Series in the MCF-7 Breast Carcinoma Cell Line

Based on our data, it is probable that a guiding principle of estrogen action is to activate several membrane-initiated kinase cascades in a single neuronal cell type. Is this borne out by data in other cell types? Addition of estrogens results in proliferation in breast cancer, typified by the response in the MCF-7 cell line. MCF-7 cells also respond to estrogens with rapid activation of the MAPK/ERK pathway via incompletely understood mechanisms. A number of studies point to different modes of MAPK activation by estrogens. For example, addition of an antibody to the heregulin receptor or depletion of heregulin from the conditioned media of MCF-7 attenuates estrogen-mediated ERK activation, suggesting that estrogen induction of growth factors plays an autocrine/paracrine role in ERK activation and hence, cell proliferation. In addition, inhibition of PKCδ and Ras signalling also blocks MAPK activation, implying an upstream role for these molecules. In tamoxifen-resistant cells with equivalent levels of ERα as wild-type MCF-7 cells, the ERK activation appears to be mediated via TGFα and increased levels of epidermal growth factor (EGF) receptors.

Liganded ER can also activate MAPK in MCF-7 cells via a rise in calcium from intracellular calcium stores. Because growth factor signalling to MAPK pathways involves the adaptor protein Shc, a similar pathway was proposed and demonstrated for estrogenic activation of MAPK. Shc binding to liganded ERα in the cytosol of breast cancer cells leads to association with the IGF receptor and subsequent activation of MAPK via the Grb2-Sos-Ras-Raf pathway. Physical association with the p85 subunit of PI3K as well as with Src has also been demonstrated with the ERα in MCF-7 cells; inhibition of PI3K activity using the inhibitor LY294002 abrogates estrogen-induced Akt/PKB and Src activation.

The transcriptional up-regulation of macrophage inhibitory cytokine-1 by PKB prolongs estrogen-induced phosphorylation of ERK. Membrane-limited E2-BSA activates PKC rapidly in MCF-7, independently of ERα or ERβ, and recurrent tumours have higher levels of PKC. Furthermore, in MCF-7 cells, estrogen-induced PI3K activates the atypical PKC, which in turn activates the src/Ras/MAPK signalling cascade. Other PKC isoforms have also been associated with proliferation in MCF-7 cells. For example, overexpression of PKCδ allows tamoxifen-sensitive MCF-7 lines to acquire tamoxifen resistance. In contrast, stable transfection of PKCα is accompanied by ERα down-regulation and elevated basal AP-1 activity. An early study also reported that transfection of this cell line with PKA or treatment with 8-bromo-cAMP could increase the agonist activity of tamoxifen on a consensus ERE-based reporter construct to about 20–75% of the level obtained with 17β-estradiol. Conversely, melanin inhibited ERα transactivation by inhibiting forskolin-induced and estrogen-induced cAMP increases.

How can these data be resolved? First, these data imply that several kinases, *i.e.*, MAPK, PI3K, PKA, and PKC, can be activated rapidly by estrogens, possibly via the ERα, in MCF-7 cells. This is similar to the data seen with neuroblastoma cells. Hence, it is probable that several kinases coordinately influence gene regulation in several target cell types to achieve a final phenotype. Finally, how proliferation of the MCF-7 cell line, *i.e.*, the final phenotype, is affected by the membrane-initiated activation of kinases is still unclear. Long-term estrogen-deprived MCF-7 cells (the LTED lines), which mimic tamoxifen resistance, show hyperphosphorylation of ERK, whereas both MAPK/ERK kinase (MEK) and PI3K inhibitors could decrease proliferation.

However, PI3K and MEK inhibitors do not decrease c-myc and c-fos expression although hyperphosphorylated retinoblastoma and cyclin D1 proteins decrease, suggesting that rapid signalling cascades are permissive to cell cycle progression but not to mitogenesis via the immediate early genes myc and fos. These studies strongly imply that estrogen-regulated rapid signal transduction cascades, that is, membrane-initiated actions are relevant to the final outcome in breast cancer cells, *i.e.*, proliferation.

We note that for full potentiation of a transcriptional effect of estrogens by a membrane-initiated action, the data using kinase activators suggest that more than one signal transduction system is sufficient, but the data with kinase inhibitors also imply that more than one signal transduction system is necessary.

How can this be? If a particular kinase cascade is "sufficient," how can another one be "necessary"? If several signal transduction systems in parallel maintain a basal state of transcriptional readiness, ablating any one of them lowers that basal state such that a genomic action of 17β-estradiol is ineffective in potentiating transcription. Conversely, if any one of the transduction systems involved is highly activated, it can substitute for the membrane-initiated action of 17β-estradiol in potentiating the later, genomic action. This idea, therefore, postulates the existence of a transcription threshold for genes.

Convergence of the Membrane-Initiated Actions of Hormones

Most early studies have treated membrane-initiated actions as disparate from the genomic actions of hormones encompassing and culminating at kinase activation or calcium increases, both of which have cytosolic substrates. Other studies, have examined the effect of growth factor receptor ligands, notably IGF and EGF, on ligand-independent activation of the ER. Such activation is particularly important in hormone therapy-resistant cancers; combined targeting to both the ER and growth factor pathways, therefore, has therapeutic significance. Another subset of studies has involved the role of the classical ERs in rapid, membrane-initiated actions, particularly in breast cancer and cardiovascular physiology. However, how the ligand at the membrane can regulate transcription in the nucleus is unclear if the same ligand, *i.e.*, 17β-estradiol has the ability to signal at both membrane and nucleus and if membrane-initiated rapid actions by the ligand influence transcription on EREs by ligand-dependent nuclear receptor action. For estrogen-regulated genes, one can mechanistically envision such coupling from membrane to nucleus at various intersecting levels such as 1) stability of the ER (protein-protein interactions); 2) nuclear translocation of the ER; and 3) modification of proteins important for transcription. Some of these aspects, especially for genes that do not contain EREs in their promoters, have been recently reviewed by Bjornstrom and Sjoberg.

Coupling to Transcription Via Stabilisation or Dimerization of Molecules Relevant to Transcription

Heat shock proteins (hsp 70/90) are important for proper folding, maturation, and translocation of the steroid hormone receptors. There are a few studies that report estrogenic regulation of the heat shock chaperones. 17β-Estradiol and bisphenol A act via the ER in uterine tissue to increase hsp90 and hsp72 mRNA via PKC. In the CNS, estrogens and androgens protect neurons from β-amyloid toxicity by increasing the levels of hsp70; microinjection of hsp70 blocked the toxicity of added intracellular β-amyloid peptide, suggesting a role in neuroprotection. However, it is not clear whether 17β-estradiol can regulate these proteins to enhance transcription driven by the ER.

The stability of the ERα has important consequences for function. The autologous down-regulation of ERα and subsequent proteolysis by the 26S proteasome is thought to be required for efficient transactivation from the consensus ERE. In MCF-7 cells, PKC can enhance proteasome degradation of ligand bound and ligand free ERα, whereas PKA, PI3K, and MAPK impede degradation. On the other hand, in pituitary tumour cells, PKA increases transactivation mediated by 17β-estradiol-liganded ERα from a consensus ERE while protecting the ERα from estrogen-induced degradation. This suggests that 17β-estradiol increases transcription, partly by autologous down-regulation of the ERα, whereas PKA could increase transcription by stabilising the ERα. The relevance of ERα degradation is, however, controversial. For example, proteolysis is not required for ERα-mediated induction of the prolactin gene or for proliferation in anterior pituitary lactotrope cells, demonstrating different requirements for the stability of ERα, depending on the ERE.

Coupling to Transcription Via the Physical Movement of Molecules into the Nucleus

17β-Estradiol can also regulate the translocation of the ER, possibly via kinase cascades. PKCδ activation, via GSK-3 results in nuclear translocation of the ERα with a concomitant increase in transcription. In endometrial adenocarcinoma, estrogenic up-regulation of p38MAPK, in turn, increases ERα phosphorylation on Thr^{311}; mutation of this site inhibits nuclear localisation as well as the interaction with coactivators. In human vascular smooth muscle cells, the MAPK phosphorylation of the ERα on Ser^{118} and nuclear translocation could also be ligand-independent.

Coupling to Transcription Via Phosphorylation of Molecules Relevant to Transcription

Our inhibitor studies strongly suggest that kinases such as PKA, PKC, PI3K, and MEK are activated during rapid exposure to either membrane-impermeable estrogen (E2-BSA) or T_3 at the membrane in neuroblastoma cells. Phosphorylation of molecules relevant to transcription, for example, the ERα itself, by these kinases could provide a simple coupling mechanism between events at the membrane and transcriptional events in the nucleus. Endogenous ER has a basal level of phosphorylation that is enhanced by physiological concentrations of estrogen at amino acid Tyr^{537} and at serines at positions 104, 106, 118, 154, and 167 in the activation function (AF)-1 N-terminal domain. Estrogen-induced phosphorylation occurs via kinase activation. PKA can phosphorylate the ER at serines 236, 305, 338, and S518. In contrast, PKC phosphorylation of the ER may be at Ser^{118} in the AF-1 domain. PI3K-activated PKB has been shown to phosphorylate ERα on Ser^{167} *in vitro*. MAPK phosphorylation of ERα occurs on Ser^{118} *in vivo* as well as *in vitro*. How phosphorylation affects ERα function is less clear. PKA phosphorylation at Ser^{236} in the DNA binding domain increases dimerization of the ER. A second PKA phosphorylation site at Ser^{305} in the ERα blocks acetylation of the ERα at the proximal Lys^{303} site, enhancing transcriptional ability. Mutations at the Ser^{118} and Ser^{167} sites decrease transcription in fibroblast and epithelial cells, suggesting that these sites were critical for transcription. *In vivo*, MAPK-mediated phosphorylation of Ser^{167} occurs via the ribosomal S6 kinase, whereas *in vitro*, Ser^{167} is phosphorylated by both Akt and casein kinase II. Phosphorylation on Ser^{305} also increases transcription from the cyclin D1 gene in breast cancer. This is possibly because phosphorylation of nuclear receptors appears to enhance interaction with coactivators. In support of this idea, Ser^{118} phosphorylation in the AF-1 domain facilitates the interaction of the ERβ with the steroid receptor coactivator 1 (SRC-1), hence increasing the transcription rate. Translocation of the receptor may also be affected by kinase-induced phosphorylation. Blocking MAPK causes loss of phosphorylation on Ser^{294} and nuclear accumulation of PR-B and prevents ligand-dependent degradation of the receptor. In turn, this decreases transcriptional activation.

Phosphorylation Sites on ERα and the Impact of Phosphorylation on Function

Other molecules may also be kinase targets. In COS-1 cells, PKA and subsequent MAPK activation results in an enhanced phosphorylation of SRC-1 leading to an increased transactivation by the chicken PR. MAPK activation of amplified in breast cancer (AIB1), a coactivator enhanced in breast cancer, stimulates the recruitment of p300 and increased estrogen-dependent ER transactivation. Furthermore, Bert O'Malley's group has provided evidence that various kinases differentially phosphorylate SRC-3/AIB1, leading to selective binding to downstream transcription factors. In MCF-7 cells, Ishikawa cells, and HEK293 cells, p38 MAPK stimulated transcription by both ERα and ERβ, leading to cell proliferation. The p38MAP stimulated transcription by the phosphorylation of a p160 coactivator, GRIP1. In addition, rapid 17β-estradiol treatment could induce the phosphorylation of SRC-3 by direct interaction between the ERα and SRC-3 outside the nucleus. Hence, the combination of a specific coactivator and another transcription factor could depend on the phosphorylation state of the coactivator molecule itself.

As is evident, none of these processes need to be mutually exclusive. Extracellular signals (exemplified in our paradigm by E2-BSA) may initially converge on a molecule, for example, the ERα, using multiple molecular mechanisms such as translocation, phosphorylation, and stabilisation. This initial convergence may lead to molecular activation, which may, in turn, impact divergent molecular processes within cells to achieve a final differentiated response to the signal. Conceivably, in neurons, the "activated" ER could enhance transcription, excitability due to effects on ion channels, and neurotransmitter release. Alternatively, the extracellular signal may activate several molecules, all of which converge on a single molecular process. For example, in our paradigm, if phosphorylation activated downstream transcription factors such as CREB binding protein (CBP), coactivators such as SRC-1 and SRC-3, and the ERα, transcription would be greatly enhanced, and amplification of the original extracellular signal would be achieved.

Initiation of Membrane-Initiated Actions at the Membrane Involves Both Classical "Nuclear Receptors" and other Proteins

The Membrane Estrogen Receptor Could be a Novel Protein in some Cell Types

The elucidation of the mER has generated enormous interest and effort because knowledge of a mER can lead to dissection of membrane-initiated signal transduction pathways activated by estrogens. Any protein that is a candidate mER should bind 17β-estradiol with specificity and saturability and be present at the plasma membrane of the cell to initiate rapid membrane-initiated action. Over the years, there have been conflicting reports on the mER in different cell types, predominantly based on the application of ER antagonists, such as ICI and immunocytochemistry. ICI, a pure ERα and ERβ antagonist, was unable to block the ability of E2-BSA to activate PKC rapidly in chondrocytes or MAPK in rat hippocampus, suggesting a novel membrane receptor for estrogen.

In the neocortex, a novel, high-affinity mER, called ER-X, which supports MAPK phosphorylation by estrogens, has been reported. A novel $G\alpha_q$ protein-coupled mER agonist, STX, can attenuate the weight gain seen in hypoestrogenic animals. This is similar to the attenuation seen with 17β-estradiol and occurs in double ($ER\alpha^{-/-}$, $ER\beta^{-/-}$) knockout mice, suggesting that a novel protein may be present at the membrane in the CNS. The identity of this receptor and its location at the membrane remains to be elucidated.

This group has also previously used an ERα/β antagonist, ICI 164,384, which blocked the ability of 17β-estradiol to rapidly uncouple the ¼-opioid receptor from an inwardly rectifying K^+ channel; however, in this study, ICI 164,384 could not block the same response induced by STX. This suggests that the depolarisation induced by 17β-estradiol and by STX has differential requirements for the presence of the classical ER. In pancreatic β-cells, binding of the membrane limited E2-HRP to the membrane can be competed by dopamine and norepinephrine, suggesting a common catecholaminergic receptor. However, the ability of norepinephrine to induce Ca^{2+} oscillations in these cells was unchanged by the addition of 100 nM17β-estradiol, which should saturate the common membrane receptor, if any.

Hence, the functional relevance of such a receptor is unclear. A novel mER has been suggested on the basis of competition binding in the GT1–7 cell line, wherein the binding of E2-HRP conjugate was competed by 17β-estradiol but not by ICI 182,780 or 4-hydroxytamoxifen.

The rapid reduction by 17β-estradiol of the Ca^{2+} transients induced by acetylcholine in these cells may involve a protein kinase G phosphorylation of the inositol triphosphate receptor that modifies Ca^{2+}release from the endoplasmic reticulum. However, the GT1–7 cell line, has been shown by others to possess the classical ERα at membrane sites. It is unclear how these conflicting data can be resolved. Is it possible that membrane-initiated signalling by 17β-estradiol in the same cell line may involve more than one type of mER?

In fish, where membrane-initiated actions of progesterone are required for oocyte maturation, the membrane PR is a novel protein coupling to G_i/G_o proteins. Recently, an orphan GPCR, GPR30, has been identified as a membrane ER in SKBR3 breast carcinoma cells based on high-affinity binding, membrane localisation, and modest levels of adenylyl cyclase activation upon estrogen addition, showing that the GPR30 can couple to $G\alpha_s$.

In SKBR3 cells that express no classical ER, GPR30 is absolutely required for 17β-estradiol activation of adenylyl cyclase. This is the first report of a mER structurally unrelated to the classical ER. In addition, binding of ICI compounds, 17β-estradiol and 17α-estradiol, occurs with equal affinity to the GPR30. On the other hand, Revankar *et al.* have reported that the GPR30 protein is uniquely localised to the endoplasmic reticulum and may serve as an intracellular ER that increases calcium in response to 17β-estradiol. However, they have also shown that GPR30-transfected SKBR3 cells can increase PIP2 concentrations, supporting the idea that GPR30 can increase nongenomic signalling in response to 17β-estradiol. However, another laboratory has reported that ERK activation and cAMP accumulation in response to 17β-estradiol is abrogated in $ER\alpha^-/GPR30^+$ SKBR3 cells. This may be due to differences in the parametres (fluorescence localisation *vs.* enzymatic activity) measured in these two studies. The localisation of GPR30 and its functional relevance in other cell types, perhaps using a selective, recently synthesized agonist, G-1, remains to be elucidated.

The Membrane Estrogen Receptor is the Classical ER in other Cell Types

Several reports support the notion that a classical ER may be tethered to the membrane. The E2-BSA activation of MAPK in Rat-2 fibroblasts and endothelial cells and the interaction between ERα and $G\alpha_i$ in IC-21 macrophages can be blocked by ICI compounds, which antagonise both ERα and ERβ. ICI 182,780 can also block CaMKII activation in the hippocampus by EB.

In astrocytes, E2-BSA can cause rapid calcium flux that can be blocked by ICI 182,780. The existence of ER at the membrane has also been inferred from functional signalling studies, combined with inhibitor studies. In immortalised GT-1 cells, picomolar concentrations of 17β-estradiol could inhibit cAMP levels, whereas nanomolar concentrations of 17β-estradiol could augment cAMP levels. In addition, these effects could be blocked by ICI 182,780 and pertussis toxin, implying that the ER was coupled to a $G\alpha_i$ protein in GT-1 cells. Qiu *et al.* have, on the basis of phospholipase C signalling by 17β-estradiol in POMC and dopamine neurons, proposed a model whereby a $G\alpha_q$ coupled receptor at the plasma membrane may rapidly activate signalling cascades in series.

Signalling in these cells was also blocked by ICI 182,780. Functionally, the lack of ERα or ERβ in double ER knockout mice abrogated the rapid ERK phosphorylation by estrogens in the medial preoptic nucleus, suggesting that the ER is required for rapid, nongenomic actions. Also, endothelial cells from double ER knockout mice, compared with wild-type mice, did not support PI3K, cAMP, and ERK phosphorylation or bind 17β-estradiol. These studies demonstrate that classical ERs that mediate genomic actions may also mediate nongenomic actions, but they do not mean that the ER is at the plasma membrane.

Using immunocytochemistry, numerous studies have shown that ERα is at the plasma membrane. In Chinese hamster ovary (CHO) cells, overexpressed ERα gives rise to a single transcript with functional protein expression in both nuclear and membrane fractions. Endothelial cells, a major estrogen target cell with relevance to cardiovascular protection, from the $ER\alpha^{-/-}$,$ER\beta^{-/-}$ mouse do not show membrane ERα. Anti-ERα antibodies on fetal rat hippocampal neurons reveal abundant staining at the neurites by confocal microscopy. In rat pituitary tumor cells, antisense oligodeoxynucleotide to ERα mRNA reduces the staining obtained with specific, high-affinity antibodies at the membrane of fixed, nonpermeabilised cells. In addition, plasma membrane fractions from MCF-7 cells that express the ERα show that the protein at the membrane that binds 17β-estradiol is the classical ERα.

How do nuclear receptors attach to the plasma membrane despite a lack of standard transmembrane structure and hydrophobicity? ERα can physically interact with membrane proteins, allowing anchorage to the membrane or to the vicinity of the membrane. Furthermore, the ER may also be present in specialised plasma membrane structures such as caveolae and lipid rafts that possess several signalling enzymes to form a organised signalling module coupled to $G\alpha_i$. Indeed, data obtained using E2-BSA do not rule out the possibility that this compound can enter and bind ERα in caveolae. Western blotting and mass spectroscopy show caveolin-1 in isolated oligodendrocyte plasma membrane colocalising with an ERβ-like molecule.

In endothelial cells, ERα associates with the p85 subunit of PI3K in caveolae, mediating PI3K/Akt activation of nitric oxide synthase (NOS). Similarly, ERβ in caveolae activates NOS, possibly via the rapid activation of MAPK. Localisation of ERα to the membrane appears to be dependent on Ser^{522}; mutation of this residue abrogates membrane localisation in caveolae in CHO cells and adenylyl cyclase activation but supports transcription, suggesting that the amino acid residues required for membrane-initiated action and genomic action may be different. In breast cancer and endothelial cells, ERα exists at the membrane as dimmers, in association with caveolin-1 in caveolae, and can support adenylyl cyclase and PI3K signalling.

NOS activation in endothelial cells was, however, not dependent on dimerization because an ERα deficient in dimerization (ERαL511R) could still activate NOS. Moreover, despite presence at the plasma membrane, nuclear localisation signal mutants could not activate ERK, whereas DNA binding mutants of the ERα could not activate ligand-dependent PI3K binding to ERα. Hence, the functional relevance of membrane localisation to signalling in endothelial cells is not clear. In addition, palmitoylation supports ERα attachment to the membrane and interaction with caveolin in HeLa and HepG2 cells, whereas addition of ligand decreases interaction of the receptor with caveolin, suggesting a classical desensitisation mechanism. In MCF-7 cells, ERα can physically bind the p85 subunit of PI3K at the membrane as well as the Shc protein via the AF-1 domain and can drive the formation of a Shc-Grb2-SOS signalling complex. In addition, a novel scaffold protein called MNAR (modulator of nongenomic action of ER) can interact with c-src using its PXXP motif while interacting with the ER using the LXXXL motifs, demonstrating an "adapter" function. Using an adapter protein to anchor at the membrane does not, however, make the ERα an intrinsic membrane protein. However, work from Cheryl Watson's group has implicated ERα at the membrane in MCF-7 cells. Using immunopanning with specific ERα antibodies, MCF-7 cells were separated by flow cytometry into high mERα and low mERα populations.

Immunopanning suggests that the antibody binding site on the ERα is available to the antibody at the surface of cells. It is difficult to reconcile this data to the previously stated data on MCF-7 cells, which show tethering of ERα to the inner plasma membrane.

One possibility is that a very small proportion of cells may have small amounts of ERα at the surface, whereas a greater number of cells in a given cell type may have ERα tethered to the membrane, although the basis of this differential distribution would be difficult to rationalise. Apart from the idea that a small fraction of endogenous cellular ERα is permanently anchored to the membrane, a small pool of nuclear ER may dynamically shuttle to the membrane.

This possibility has been demonstrated by an ERα artificially tethered to the plasma membrane with a rhodopsin fusion protein that can shuttle to the nucleus to increase transcription in 293 cells. In MDA-MB-231 breast cells stably transfected with either wild-type ERα or myristoylated ERα lacking the nuclear localisation signal, the myristoylated ERα shows membrane localisation but is not down-regulated by 17β-estradiol, as is the case for wild-type ERα. In addition, compared with the WT ERα, the myristoylated protein cannot up-regulate the transcription of estrogen reporter genes on exposure to 17β-estradiol.

Functionally, the presence of ER at the membrane may have physiological consequences that need to be studied in more detail. In MCF-7 variants sorted for higher ERα localisation to the membrane, estrogenic induction of adenylyl cyclase is correlated with lower proliferation. In *Xenopus laevis* oocytes, MNAR contributes to meiotic arrest because reduction of MNAR mRNA increased liganded androgen receptor-driven MAPK activation and oocyte maturation. In the CNS, differentiation of midbrain dopaminergic neurons appears to be due to membrane-attached ERα in astrocytes.

Do Different Cell Types have Different Receptors?

The wealth of studies demonstrating novel proteins as the mER or the classical nuclear receptor, the ERα, as a candidate mER suggest that different cell types may possess different mERs. None of the novel proteins that are mER candidates (*i.e.*, ER-X or GPR30) show any homology to the novel membrane PR, cloned from sea trout ovaries, indicating that, unlike the classical nuclear receptor superfamily, there may not be a parallel "membrane receptor superfamily". It is also possible that different molecules may "moonlight" as the mER either temporally or for distinct functional roles in a single cell type. Additionally, the domains required for ER function at the membrane could be cell specific, depending on the molecule that the ER may be tethered to in that cell type. These ideas about the nature of membrane receptors for hormones remain to be clarified.

Future Directions

Data from other laboratories and ours suggest that both kinases and calcium flux appear to be important in E2-BSA-mediated actions on transcription in neurons. Several different, parallel pathways can couple signal transduction cascades to transcription in the nucleus. For example, ligand-independent transcription (which is not the focus of this review) results from activation of growth factor receptors at the membrane and subsequent ERK activation. In addition, binding of cognate ligands to membrane receptors such as the mER or a novel protein that acts as the mER can activate kinase cascades which, in turn, can facilitate transcription from either EREs or other enhancer elements.

Because initiation of kinase activation from the plasma membrane by estrogens is an emerging story, several questions remain. E2-BSA and E2-HRP have been the only membrane-limited probes used until the synthesis of an estrogen-dendrimer conjugate. Their specificity has not always been demonstrated in all studies, and their uses *in vivo* have been particularly limited. The use of the estrogen-dendrimer conjugate, synthesized in Dr. John Katzenellenbogen's laboratory, should help to clarify membrane-limited effects.

Are several kinases activated rapidly by membrane-limited estrogens to regulate transcription from a coordinate set of genes? The transcription of this set of genes would then change the physiology of the cell in response to the stimulus at the membrane. The next step would be to elucidate all kinase and other rapid signal transduction pathways that contribute to transcription and investigate whether they are absolutely necessary to transcriptional regulation. Most studies tend to look at a single kinase or a single rapid signal transduction pathway without considering the crosstalk that is inherent in these pathways. Detailed pathway profiling in response to brief administration of estrogens that is relevant to transcription as well as a physiological end point, *e.g.*, reproduction or neuroprotection, is the next step in these studies. Models of rapid signalling pathways could then be generated for each physiological scenario affected by estrogens.

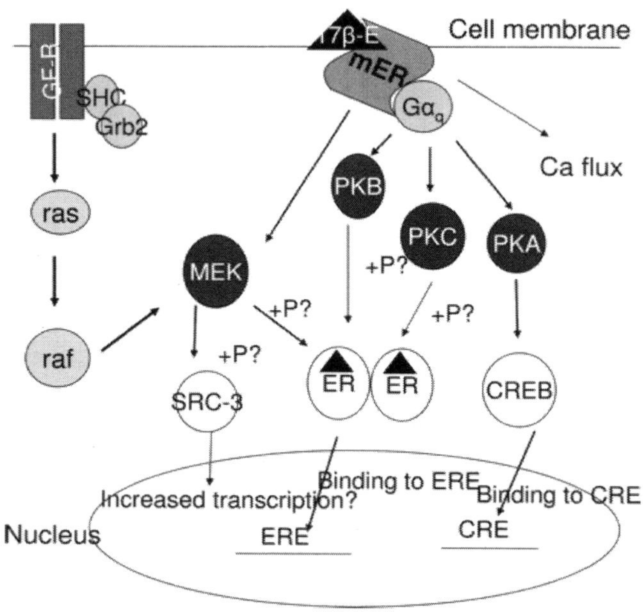

Figure: Parallel pathways that integrate signals at the membrane to transcription in the nucleus in neuroblastoma cells. 17b-Estradiol (Δ) binds to a $G\alpha_q$-coupled membrane ER. This initiates several kinase cascades (PKA, PKB, MEK, PKC) and Ca^{2+} flux. Kinases modify the ER bound to the ligand or another transcriptionally relevant molecule, such as the coactivators, SRC-3. The modified ER can translocate to the nucleus where it enhances transcription via the ERE in gene promoters. In contrast, the ER can also be activated by growth factors (ligand-independent mechanisms) via the MEK pathway. In addition, the activation of kinase cascades by membrane-limited estrogens can also result in increased transcription from non-ERE containing promoters. In this figure, a promoter with a CRE is shown; this can be bound by CREB protein, which in turn is activated by PKA. GF-R, Growth-factor receptor.

What are the molecular mechanisms by which membrane-initiated actions influence transcription for cognate ligands? Is there a point of convergence for kinase cascades? Mechanistic studies on how rapid signal transduction pathways affect estrogen-driven physiology are the next step. Phosphorylation of the ER or other key transcriptionally relevant molecules could be key coupling events and act as a point of convergence for membrane-initiated kinase cascades and transcription in the nucleus. Recently, O'Malley's group showed that the phosphorylation of SRC-3 by 17β-estradiol can occur in the presence of a transcriptionally inactive ERα in the cytoplasm.

This suggests that SRC-3 can be phosphorylated in an extranuclear location by interaction with ERα. Another critical area for study is the importance of phosphorylation of transcriptionally relevant molecules. For the ERα, for example, there is little consensus on the effects of phosphorylation on transcription, and this needs to be studied in detail. SRC-3 phosphorylation occurs on six different sites; stimulus specificity could be encoded by different combinations of phosphorylation sites.

Phosphorylation of nontranscriptional targets, *e.g.*, ion channels or intracellular Ca^{2+} receptors, could also represent a point of convergence of rapid signalling cascades. Apart from phosphorylation, other processes such as stabilisation of transcriptionally relevant molecules, methylation of promoters, or control of proteolytic degradation could also be important. The relative contributions of these molecular targets to the final phenotype are ripe for exploration in the context of nuclear receptor action in the CNS.

What are all the various physiological processes that can be influenced by these molecular mechanisms? Much of the classical studies that involve membrane actions of estrogens in the CNS have been inspired by estrogen's control of reproductive physiology and behaviour. Which genes that are relevant for lordosis behaviour in rodents or other nonreproductive functions are impacted by the membrane-initiated actions of estrogens? Uncovering these genes will be important to understand behaviours driven by estrogen.

Because membrane-initiated signal transduction events can potentiate transcription by estrogens, it is obvious that other extracellular signals that impact these pathways may have an effect of transcription. Crosstalk between growth factor receptors and the ER has been studied and reviewed extensively, especially in the context of cell proliferation in cancers. In the neuroendocrine context,

EGF has been shown to selectively activate ERα-dependent, but not PR-dependent lordosis behaviour in a ligand-independent fashion. In addition, dopamine receptor D1 agonists increase lordosis behaviour in Sprague Dawley rats; this effect is dependent on intact D5 receptors in the VMH and the PR. How specificity is maintained during such ligand-independent crosstalk remains to be studied. In neuroblastoma cells, T_3, the genomically more active form of the hormone, via thyroid hormone receptor (TR) α1 or TRβ1 potentiates ERα-mediated transcription from a consensus ERE-driven luciferase reporter compared with a single 17β-estradiol pulse, similar to the potentiation caused by E2-BSA, via activation of MAPK signalling. Phosphorylation of ERα on serine residues 118 and 167 and the presence of 17β-estradiol is also important in T_3-mediated transcriptional potentiation. This is an example of ligand-dependent crosstalk between estrogens and thyroid hormone that may play roles in mood and cognition); this remains to be elucidated in a physiological scenario.

The novel idea that genomic transcription by hormones, *i.e.*, ligand-dependent transcription at hormone response elements, can be affected by membrane-initiated signal transduction events initiated by cognate or noncognate ligands is a paradigm shift in nuclear receptor biology.

Bibliography

Anderson, A. Trevor: *Fish Nutrition in Aquaculture*, Springer, Delhi, 2009.

Bailey, MT: *A-Z of Tropical Fish Diseases and Health Problems*, Howell Book House, New York, 2001.

Balfour, F. M.: *A Monograph on the Development of Elasmobranch Fishes*: Macmillan, London, 1878.

Barlow, S.M.L: *Chemical and Biological Assay Procedures for Lysine in Fish Meals*, J.Sci.Food Agric., 1984.

Basant, Puri K.: *The Natural Way to Beat Depression: the groundbreaking discovery of EPA to change your Life,* Hodder and Stoughton, London, 2004.

Boserup, E.: *The Conditions of Agricultural Growth: The Economics of Agrarian Change under Population Pressure*, Aldine Press, Chicago, 1965.

Bowman, S.J.: *Digest of Statistics of IAFMM*, Potters Bar, U.K., IAFMM, 1984.

Bruno D.W., Poppe T.T., *A Color Atlas of Salmonid Diseases*, Academic Press, London, 1996.

Cech, JJ and Moyle, PB: *Fishes, An Introduction to Ichthyology*, Benjamin Cummings, NY, 2003.

Chow, Ching Kuang: *Fatty Acids in Foods and Their Health Implications*, Routledge Publishing, New York, 2001.

Clover, Charles: *The End of the Line: How Overfishing is Changing the World and What we Eat*, Ebury Press, London 2004.

Ferguson, H.W.: *Systemic Pathology of Fish*, Iowa State Press, Ames, Iowa, 1989.

Geran, James: *The Proper Care of Goldfish*, TFH Publications, NJ, 2000.

Hamm, Wolf: *Trans Fatty Acids*, Blackwell, NY, 2008.

Haskell, B.D.: *Ecosystem Health: New Goals for Environmental Management*, Island Press, Washington, D.C., 1992.

Hasler, A.D.: *Orientation and Fish Migration*, Academic Press, New York, 1971.

Kulkarni, G.K.: *Fisheries and Fish Toxicology*, APH, Delhi, 2006.

Lloyd, G.: *Mammals of the Pacific States,* Stanford University Press. 1965

Maddy Hargrove: *The Discus: An Owner's Guide to Happy Healthy Fish*, Howell Books, NY, 1999.

Mahadevan, R.K. and Selvamani, B.R.: *Fish and Fishery Culture: Assessment and Evaluation*, Campus Books International, Delhi, 2008.

Malvee, Sangeeta: *Fish Genetics*, SBS Pub, Delhi, 2008.

Martin, A.M.: *Fisheries Processing: Biotechnological Applications*, Chapman and Hall, Delhi, 2009.

Mathur, A.K. and Sharma, L.L.: *Hand Book of Freshwater Ornamental Fishes,* Yash Pub house, Delhi, 2006.

Mehta, Varun: *Fisheries and Aquaculture Biotechnology*, Campus Books International, Delhi, 2006.

Mori, Fumitashi and Sakamoto, Yohei: *Aquarium Fish of the World: The Comprehensive Guide to 650 Species*, Chronicle Books, UK, 1991.

Moyle, PB and Cech, JJ: *Fishes, An Introduction to Ichthyology*, Benjamin Cummings, NY, 2003.

Nelson, J. S.: *Fishes of the World*, John Wiley & Sons, Inc.. , NY, 2006.

Petr, Tomi: *Fisheries in Irrigation Systems of Arid Asia*, Daya, Delhi, 2007.

Pollard, D. A. *Freshwater Fishes of South-eastern Australia*. McDowell, R. M. Reed, Sydney. 1980.

Ponniah, A.G.: *Fish Biodiversity of India*, NBFGR, 2002.

Ralph, R.: *Breeding Australia's desert goby*, Tropical Fish Hobbyist, 1988.

Rao, D.V.: *Poisonous and Venomous Fishes of Andaman Islands, Bay of Bengal,* Zoological Survey of India, 2003.

Rebelin, W.E.: *The Pathology of Fishes*, The University of Wisconsin Press, UK, 1975.

Rosen, D. E.: *Modes of Reproduction in Fishes, How Fishes Breed*, Natural History Press, New York, 1966.

Sharma, Arti: *Fishes: Aid to Collection, Preservation and Identification*, Daya, Delhi, 2006.

Singh, Braj Prasad: *Fish Genetic and Biotechnology*, Swastik, Delhi, 2008.

Stoskopf, M.K.: *Fish Medicine*, W.B. Saunders Co. 1993.

Tunstall, Jeremy: *The Fisherman: The Sociology of an Extreme Occupation*, McGibbon and Kee, London, 1969.

Untergasser, D.: *Handbook of Fish Diseases*, T.F.H. Publications, Neptune, NJ, 1989.

VanStone, J. W.: *Athapaskan Adaptations: Hunters and Fishermen of the Subarctic Forests*, Aldine, Chicago, 1974.

Villars, T.: *The Chemical Senses of Fishes*, Academic Press, London, 1974.

Ward, Barbara E.: *Varieties of the Conscious Model: Fishermen of South China,* Tavistock, London, 1965.

Warner, William W.: *Beautiful Swimmers: Watermen, Crabs and the Chesapeake Bay*, Little, Brown, Boston, 1976.

Index

F

G

H

S

T

V

W

❑❑❑

Ko49
147